MW00835437

Sea Monsters

Sea Monsters

A VOYAGE AROUND THE WORLD'S MOST BEGUILING MAP

JOSEPH NIGG

THE UNIVERSITY OF CHICAGO PRESS

Chicago and London

Joseph Nigg is one of the world's leading experts on fantastical animals, and his exploration of the rich cultural lives of mythical creatures has garnered multiple awards and been translated into more than twenty languages. He is also the author of *The Book of Gryphons*, *The Book of Fabulous Beasts: A Treasury of Writings from Ancient Times to the Present*, and *How to Raise and Keep a Dragon*, among others.

The University of Chicago Press, Chicago 60637
The University of Chicago Press, Ltd., London

Printed in China

22 21 20 19 18 17 16 15 14 13 1 2 3 4 5
ISBN-13: 978-0-226-92516-5 (cloth)
ISBN-13: 978-226-92518-9 (e-book)
DOI: 10.7208/chicago/9780226925189.001.0001

A CIP record for this title is available at the Library of Congress.

♾ This paper meets the requirements of the ANSI/NISO Z39.48-1992 (Permanence of Paper).

Color origination by Ivy Press Reprographics.

Cover images: courtesy of James Ford Bell Library, University of Minnesota

This book was conceived, designed, and produced by
Ivy Press
210 High Street, Lewes
East Sussex BN7 2NS
United Kingdom
www.ivypress.co.uk

Creative Director Peter Bridgewater
Publisher Jason Hook
Editorial Director Caroline Earle
Art Director Michael Whitehead
Designer Andrew Milne
Project Editor Jamie Pumfrey

For Joey
and
In memory of my father

FINMARCHIA
SCRICFINIA
BIARMIA
LAPPIA OCCIDENTALIS
LAPPIA ORIENTALIS
MARE BOTNICV
BOTNIA OCCIDENTALIS
BOTNIA ORIENTALIS
CARELIA
ANGERMANNIA
MIDDELPADIA
SCANDIA
NORVEGIA REGNUM
SVECIA
TAVASTIA
FINLANDIA VEL FINNINGIA OLIM REGNUM
MARE SVETICVM
MARE FINONICUM SIVE SINUS VENEDICUS
MOSCOVIE PARS
RUSSIA ALBA
LIVONIA AQUILONARIS
MARE LIVOICUM
LIVONIA AUSTRALIS
GOTHIA
DANIA
MARE GOTICUM
SAMOGETHIA OLIM REGNUM
LITUANIE PARS
HOLSATHIA
POMERANIA
SAXONIA

CONTENTS

APPENDICES

Invitation To A Voyage

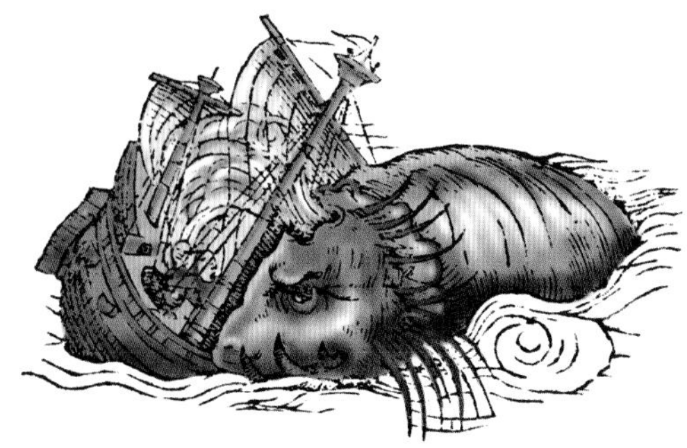

The first time one looks at a color print of Olaus Magnus's 1539 *Carta Marina*, the eyes scan the crowded land mass and fix on the creatures in the western part of the map. Larger than the other images and framed by open space, they dominate the chart visually and stir the imagination.

To us, the quaint figures rising in the northern waters of Olaus's map of Scandinavia could be illustrations in a children's book. However, given that maps chart human knowledge, that they provide glimpses of our understanding of the world at any point in time, Olaus's sixteenth-century contemporaries would have regarded the *Carta Marina* sea monsters differently than we do. When the chart was made, in the early years of the Age of Exploration, there was a lingering belief in the existence of griffins, unicorns, dragons, the phoenix, the monstrous races, and a host of other unnatural creatures. Modern science was in its infancy. Although adherents to the direct observation of nature would soon challenge hearsay and tradition and begin to classify animal life, at the time the medieval imagination was still free to shape its own forms of the natural world. The chart's giant lobster gripping a swimmer in its claws, a monster being mistaken for an island, and a mast-high serpent devouring sailors would have represented actual fears of the unknown deep.

Those and Olaus's other fanciful sea beasts are not mere decorations to fill empty spaces. Nor are they only visual metaphors for dangers lurking in the sea. Intended as representations of actual marine life, they are identified in the map's key. Most of them are also pictured and described in Olaus's commentary of the chart, *Historia de Gentibus Septentrionalibus* ("History of the Northern Peoples," 1555).

A voyage up Olaus Magnus's map promises sightings of monsters never seen before. Fearsome creatures such as the spouter rise for the first time in the northern waters of the Carta Marina. *The fantastic beast's likeness appears again in celebrated works of the age.*

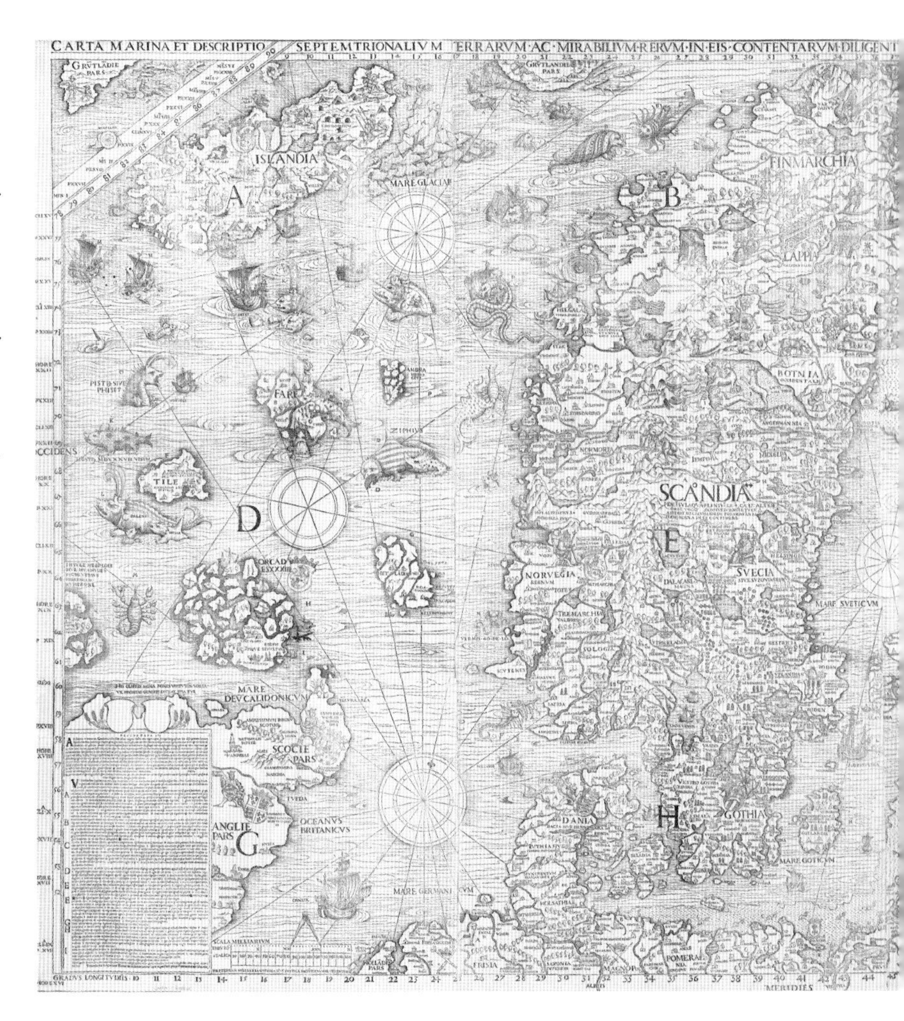

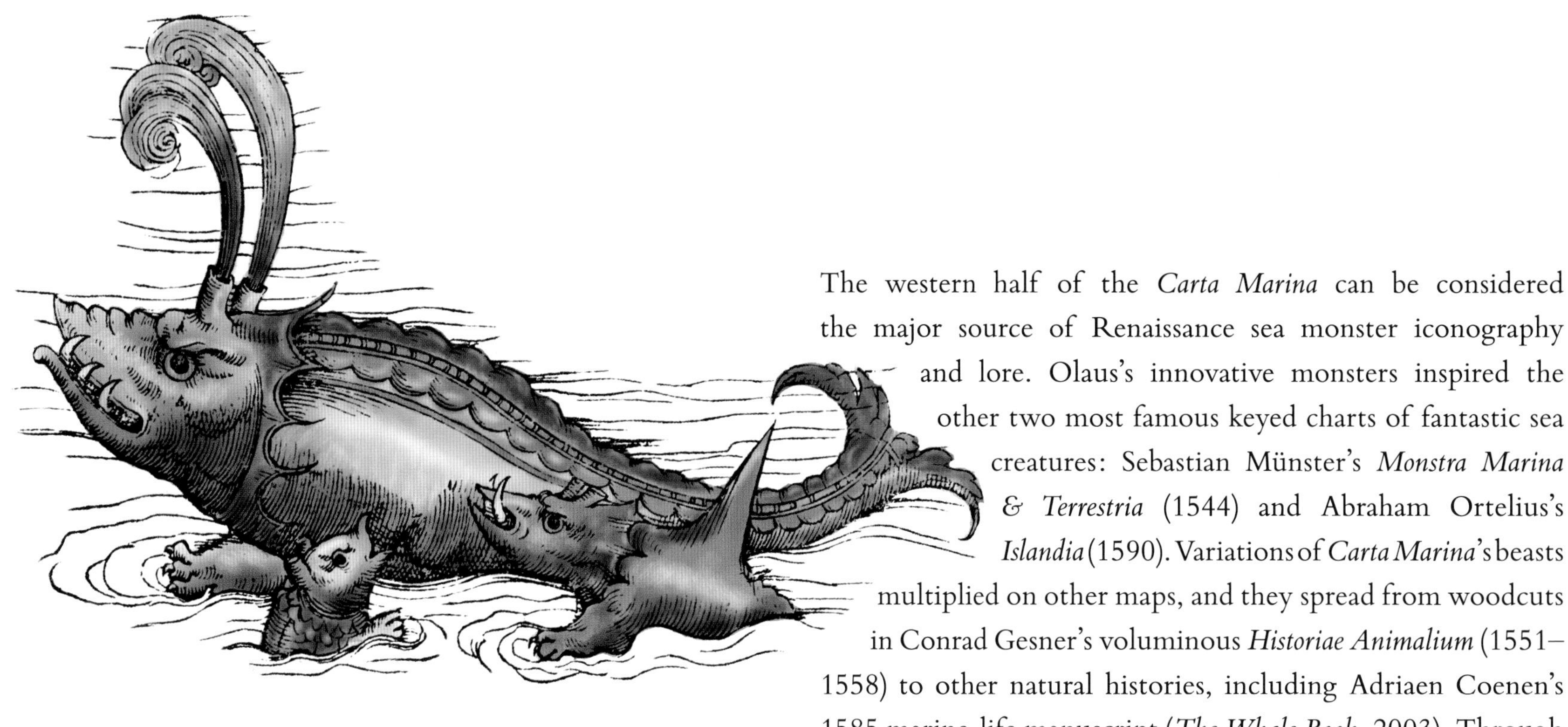

The western half of the *Carta Marina* can be considered the major source of Renaissance sea monster iconography and lore. Olaus's innovative monsters inspired the other two most famous keyed charts of fantastic sea creatures: Sebastian Münster's *Monstra Marina & Terrestria* (1544) and Abraham Ortelius's *Islandia* (1590). Variations of *Carta Marina*'s beasts multiplied on other maps, and they spread from woodcuts in Conrad Gesner's voluminous *Historiae Animalium* (1551–1558) to other natural histories, including Adriaen Coenen's 1585 marine-life manuscript (*The Whale Book*, 2003). Through his map and its voluminous commentary, Olaus became the age's principal chronicler of the sea serpent, the giant squid, and sea monsters in general. These representations were influential for centuries and are still discussed in our own time. They are the ancestors of the decorative whales that dot oceans on modern commercial globes.

A Carta Marina *mother whale nursing her calf is one of the earliest depictions of a spouter as a mammal. Elsewhere on the map, a narwhal is pictured for the first time. Both of those figures, like nearly all of Olaus's sea monsters, will be reproduced in the years to come.*

The most recently discovered of only two extant copies of Olaus Magnus's Carta Marina. *Four centuries after the 1539 printing of the wood-block map in Venice, the Uppsala University Library acquired the print from Switzerland in 1962.*

A Voyage with the Sea Creatures

To sight Olaus's beasts, this book takes the reader on an imaginary voyage up the northern seas of the *Carta Marina*, with Olaus himself as the guide. His commentary, from the first English translation of his book, *A Compendious History of the Goths, Swedes, and Vandals and Other Northern Nations* (1658), introduces each beast before it surfaces in full-blown art. The beast encounter ends with discussion of the figure's traditional lore, its legacy, and its modern forms. Reproductions of the three renowned sea monster charts, and translations of their keys, enable the reader to cross-reference influential images throughout the book. Surveys of sea beasts' mythical beginnings and natural history complete preparations for the voyage.

OLAUS MAGNUS

Perhaps the only known portrait of Olaus Magnus. Both the traveler and the horse wear snowshoes as they cross a mountain pass between Sweden and Norway. Olaus explains that the plates are like shields fastened to the feet. To Carta Marina *scholar Edward Lynam, the scene depicts an actual event. The reversed* History *vignette corresponds to the* Carta Marina *image (E f) just a little east of the Sea Cow.*

The man whose map and *History* became influential sources of sea monster lore up to our own time was a noted ecclesiastic, cartographer, and historian in his own age.

Olaus Magnus (Olaf Månsson, 1490–1557) and his elder brother, Johannes, were Catholic priests who sought exile following their native Sweden's Reformation conversion to Lutheranism. Both brothers, born in Linköping, Sweden, had traveled throughout Europe in service to the Church, and both produced nationalistic works. Johannes's earlier appointment as Archbishop of Uppsala passed on to Olaus after the brother's death—but only in name, not as an active position.

The brothers were living in Danzig (modern Gdansk), Poland, when Olaus began work on a map of the northern regions in 1527, the same year that Sweden became Protestant. The map would introduce Europeans to the rich history and culture of the peoples of formerly Catholic Scandinavia. Changing vague notions of his homeland entailed correcting a particular recent map.

The Vikings left no charts of their voyages, but a twelfth-century manuscript, *The Book of the Settlement of Iceland*, lists sailing directions and times on the Norwegian Sea: "from Norway, out of Stad, there are seven half-days' sailing to Horn, in eastern Iceland, and from Snowfells Ness, where the cut is shortest, there is four days' main west to Greenland." Figures representing Northern countries appear on the famous *Hereford Mappa Mundi* (ca. 1300) and the first printed world map, the *Rudimentum Novitiorum* (1475). Those were derived from medieval "T-O" maps that divided the circular world into Asia, Europe, and Africa, with Jerusalem in the center. It was not, however, until cartographers charted land masses from the second-century numerical coordinates of Greek geographer Ptolemy that the region began to take visual shape. The 1482 Ulm edition of Ptolemy's *Geographia* contained the first printed map of the North. Adapted from a manuscript version by Claudius Clavus, this trapezoidal map of Nicolaus Germanus remained the standard charting of the North for decades.

Isidore of Seville's seventh-century "T-O map," printed in 1472. The first diagram of the world printed in Europe, it remained in print during Olaus's time. Within the circular ocean are the continents of Asia, Europe, and Africa. The "T" in the center represents the Christian Cross. The diagram is from Etymologies, *one of Olaus's standard sources.*

The Map and the History

Dissatisfied with the Germanus map, Olaus labored for nearly twelve years to produce a more accurate one of his beloved region. In the meantime, he added a newer charting of Scandinavia from Jakob Ziegler's 1530 *Schondia*. Olaus's original wood-block map comprised nine folio sheets, the total wall map being about 4 feet high by 5 feet wide. Printed in Venice, Olaus's *Carta Marina* was the largest, most detailed, and most accurate map of any part of Europe up to that time.

Only a few copies of the expensive map were printed, but the *Carta Marina* influenced the charting of Scandinavia up to the early seventeenth century. The French-born Italian engraver and publisher Antonio Lafreri

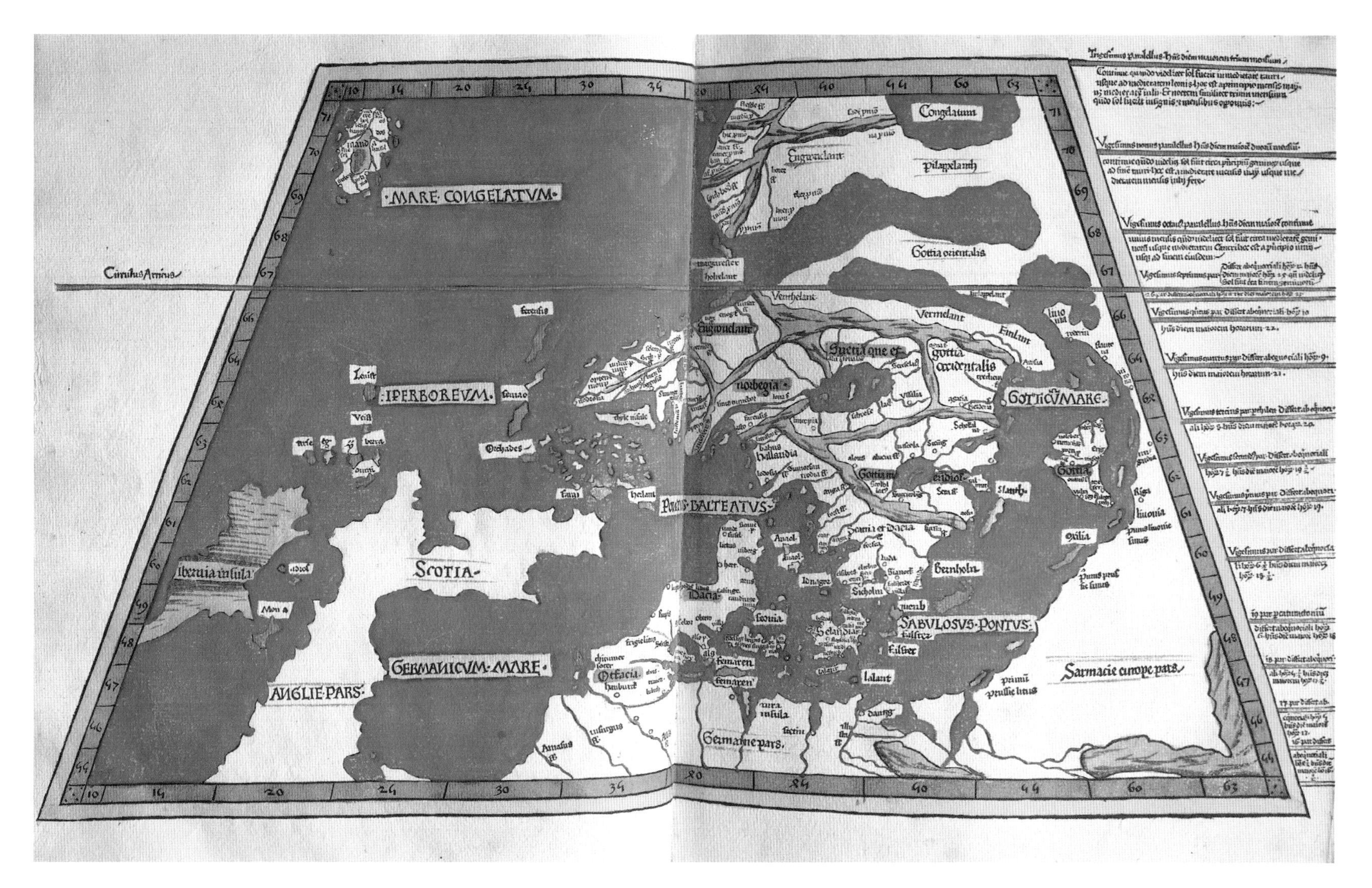

Nicolaus Germanus's 1482 map of Scandinavia, the first printed map of the North. From the Ulm edition of Ptolemy's second-century Geographia, *it is the influential trapezoidal map that Olaus set out to correct with his* Carta Marina. *The hand-colored woodcut map, with Greenland directly north of Scandinavia, is based on the 1427 manuscript map of Claudius Clavus.*

produced a smaller copy in 1572, and the fresco of Scandinavia still in the Vatican's Map Gallery was heavily influenced by Olaus's map. Nonetheless, copies of the original *Carta Marina* dropped out of circulation by the 1580s, and Italian and German translations of the Latin key, the *Opera Breve*, also became rare. It was not until 300 years later, in 1886, that a copy of the original was discovered in the Munich state library. It was thought to be the only one in existence before another surfaced in Switzerland in 1962. That second copy is now in the University of Uppsala Library in Sweden.

Olaus noted on the *Carta Marina* that he would supplement the map with a book explaining the figures. And so he did—after working on it for the next sixteen years. During that time, he and Johannes lived in Venice and moved to Rome, where Johannes died in 1544. Olaus devoted entire chapters to the map's figures in his comprehensive *History.* Sea monster chapters dominate the work's penultimate section, Book 21.

The *History* provides a blend of personal observations, Scandinavian culture, and scholastic dependence on ancient authors. It was the first comprehensive study of people of the Nordic countries. Olaus completed the work while serving as the head of the Swedish convent of St. Birgitta in Rome; he died two years later. Publication of multiple editions attests to the book's popularity into the middle of the next century.

Olaus Magnus's
CARTA MARINA

From the northern seas portion of the *Carta Marina*, reproduced here are six of the nine "A" through "I" panels of the total map—from left to right in rows across the top, center, and bottom of the chart. Each panel is labeled with a large letter. Figures within each section are marked with smaller letters and are identified in the key in the map's lower left corner. The jacket of this book folds out into a color facsimile of the entire *Carta Marina*, and a grid of the map accompanies the complete key on pages 150–151.

The entire *Carta Marina* is both a sea and land map, as declared in its full title: "A marine map and description of the northern countries ..." The bulk of the map is devoted to a charting of the region, with its myriad of pictures depicting the rich history, mores, folklore, and natural history of Scandinavia. Degrees of latitude and longitude as well as length of days are indicated in the map's frame. The common "*Carta Marina*" name, compass roses with directional rhumb lines, and a distance scale with dividers are in the tradition of navigational charts. The northern seas of beasts and images of maritime life occupy more than the western third of the map.

Sea creatures graphically dominate the mass of figures on the total *Carta Marina*. Framed by open space, they rivet the attention of the viewer by their size and fantastic forms. Several are the stuff of nightmare, the shapes of Renaissance mariners' fears. Among them are many unnamed "monsters." Unlike the map's familiar land animals, which are drawn from life, most of *Carta Marina*'s sea beast figures were born of the artist's imagination, by someone who had heard such animals described. Olaus gleaned much of his knowledge of marine life from tales told by fishermen and sailors. Standard names, drawings, and classification of species were yet to come, following systematic observation of nature.

In the decades of the waning medieval imagination, Olaus's fantastic creatures surface not only in Conrad Gesner's great, transitional natural histories, but also on the other two most famous Renaissance charts of sea monsters: Sebastian Münster's *Monstra Marina & Terrestria* and Abraham Ortelius's *Islandia*.

SEA MONSTER KEY

*Adapted from the Uppsala University Library's translation of Olaus Magnus's Latin key.**

— A —

(K) Sea monsters, as huge as mountains, capsize the ships if they are not frightened away by the sounds of trumpets or by throwing empty barrels into the sea.

(L) Seamen who anchor on the backs of the monsters in belief that they are islands often expose themselves to mortal danger.

(1) Southwest—The Sea Unicorn (not keyed on the original map).

— B —

(B) Two colossal sea monsters, one with dreadful teeth, the other with horrible horns and burning gaze—the circumference of its eye is 16 to 20 feet.

(2) Northwest—A Sea Creature (not keyed on the original map).

(C) A whale rising up and sinking a big ship.

(D) A worm 200 feet long wrapping itself around a big ship and destroying it.

(E) Rosmarus, a sea elephant, sleeps hanging from the cliff and is caught thus.

(F) Several horrendous whirlpools in the sea.

— D —

(A) The Faroe Island; its fish-eating inhabitants cut up and divide among themselves the big sea animals thrown up by the storms.

(D) The terrible sea monster Ziphius [actually "Xiphias," swordfish] devouring a seal.

(E) Another grisly monster, name unknown, lurking at its [the Ziphius's] side.

(I) Ducks being hatched from the fruit of the trees.

(K) A sea monster similar to a pig.

(L) A whale, a very great fish, and the Orca, which is smaller, his deadly enemy.

(M) A Polypus, or creature with many feet, which has a pipe on his back.

(O) The Whirlpool, or Prister, a kind of whale whose floods of waters sink the strongest ships.

(P) Spermaceti, which is called ambergris.

— E —

(E) A monster looking like a rhinoceros devours a lobster which is 12 feet long.

(M) A sea snake, 30 or 40 feet long.

(3) West—The Sea Cow (not keyed on the original map).

— H —

(F) The benevolence of the fishes called rockas in Gothic and raya in Italian: They protect the swimming man and save him from being devoured by sea monsters.

* "L" to "P" in panel D were omitted from Olaus's key—these entries are adapted from the 1658 translation of the *History*.

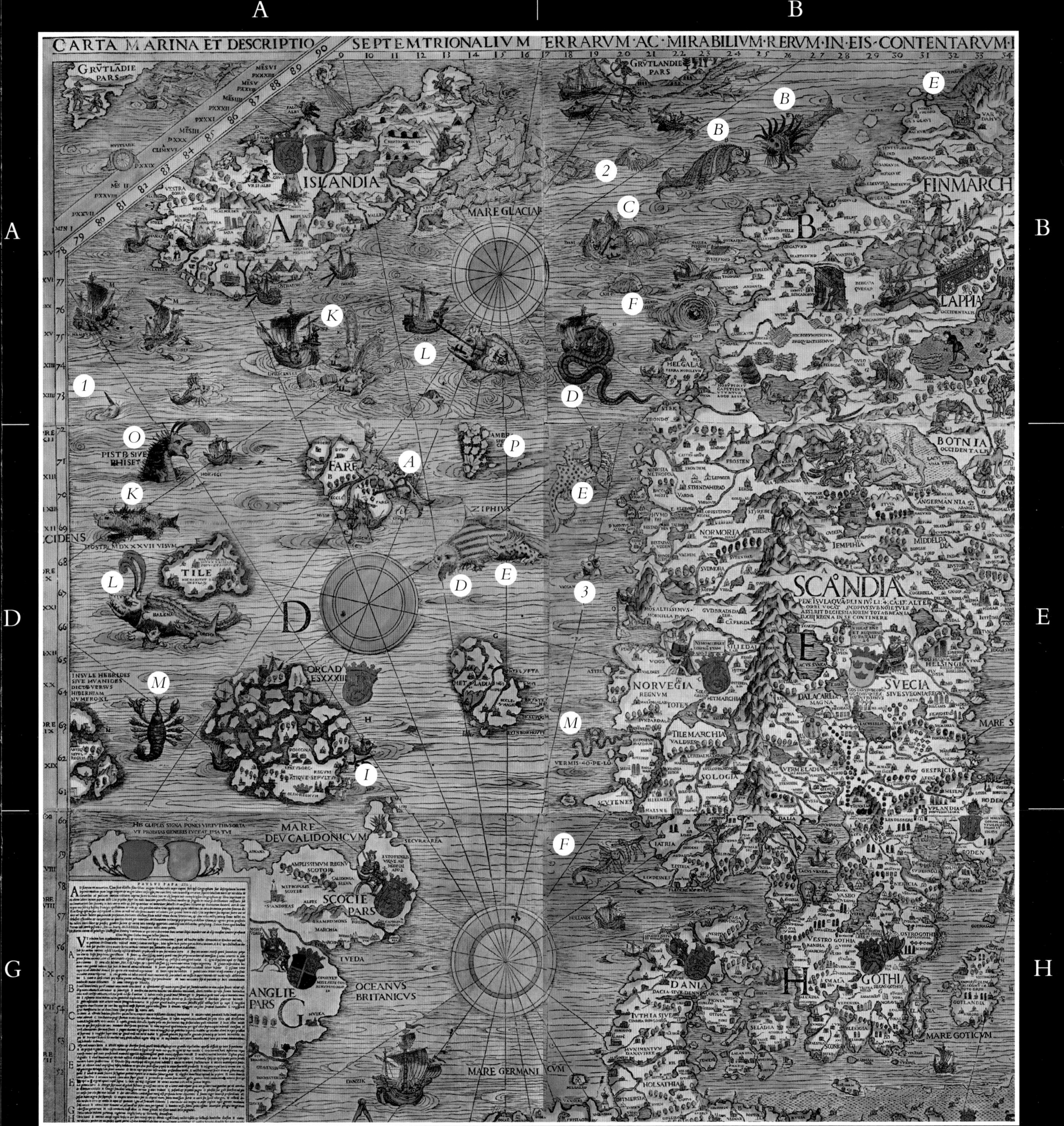

A
B
CARTA MARINA ET DESCRIPTIO SEPTEMTRIONALIVM TERRARVM AC MIRABILIVM RERVM IN EIS CONTENTARVM
GRVTLADIE PARS
GRVTLANDIE PARS
ISLANDIA
MARE GLACIAL
FINMARCH
LAPPIA
BOTNIA OCCIDENTALIS
FARE
ZIPHIVS
TILE
SCANDIA
NORVEGIA REGNVM
SVECIA
ORCADES XXXIII
VERMIS 40 PE LO
MARE DEVCALIDONICVM
SCOCIE PARS
ANGLIE PARS
OCEANVS BRITANICVS
MARE GERMANICVM
DANIA
GOTHIA
MARE GOTICVM

Sebastian Münster's

MONSTRA MARINA & TERRESTRIA

The indebtedness of Sebastian Münster (1488–1552) to Olaus's *Carta Marina* is evident in the renowned German scholar's woodcut plate of sea and land "monsters" of the North. All of Münster's animals—whether in the sea or in the land panel at the top—are derived from figures on Olaus's map. Some, such as the lobster (M) and the sea swine (K), are virtual copies of the originals; others are artistically modified. Nearly all Olaus's marine images, reversed through printing, fill the chart. And Münster's lettered key often echoes that of Olaus.

Like the sea beasts of the *Carta Marina*, those from Münster's *Cosmographia Universalis* (1544 and after) are intended to represent real animals. They comprise an illustrated catalog of imagined marine life. The tusked, spouting monster that attacks a ship at the top left (A) is pictured again—at the bottom—in its full, detailed form (A), as if it were serious natural history. The engraver's most accurately rendered "monster," as it is on the *Carta Marina*, is the gigantic lobster.

Münster was one of the preeminent cartographers of the sixteenth century. His best known works are editions of Ptolemy's *Geographia* (1540 and after) and his voluminous *Cosmographia*. His portrayals of Scandinavia in editions of both books were influenced by Olaus's wall map. The latter volume went through nearly forty editions in Latin and European languages, becoming one of the most popular and influential books of its time.

In his *History,* Olaus cites Münster's earlier *Cosmographia* several times, and even refers to him by name, completing the cross-germination between the two authors. The *History* itself would qualify as the "great book" Münster refers to in the final entry of his key.

SEA MONSTER KEY

(A) Great whales as large as mountains are sometimes seen near Iceland, which overturn ships unless frightened off by the sound of trumpets, or they distractedly play with empty barrels dropped into the sea. It also happens sometimes that sailors are put in danger when they fix anchors onto the backs of whales which they think are islands. In the language of these men they are called Trolual, that is, devil whales.

(B) A terrifying type of huge monsters, called physeteres. Solinus makes mention of these after Pliny. This whale upright submerges even a large boat, drawing up water and blowing it out through the holes of its forehead in the manner of a rainstorm.

(C) In the ocean one finds sea snakes, 200 and 300 feet long. They twist around the ship, harm the sailors, and attempt to sink it.

(D) Two large sea monsters, one fierce with teeth, the other with horns and terrifying with a flaming appearance, whose eye's circumference has the measurement of 16 or 20 feet. The head is square with a great, long beard, but the rear part is small.

(H) This horrible sea monster is called Ziphius, and it eats a seal, which they translate in German as *selehund*.

(I) Ducks are born from the fruit of certain trees in Scotland.

(K) Sea monster, somewhat resembling a pig, was seen in AD 1537.

(L) A type of whale, called orca by some, which is called the springval by the Norwegians because of its sprightliness, and it has a high, wide hump on its back.

(M) Crabs of notable size, which the people of Humeren call gambars, are of such strength that they seize men with their claws and overwhelm them.

(N) A monster like the rhinoceros devours a gambar, or crab, of 12 feet, has a horned and sharp nose even as a back sloping with a sharp point.

(R) The pelican, a bird not smaller than a duck, when its throat is full of water, sends out a sound like that of a braying donkey. It has a lump like a sack under its beak.

(S) The mercy is noted of fish which are commonly called Rocken and in Italian Raya for they save and watch over a submerged man lest he be devoured by sea monsters.

(T) That monster is called a sea cow because it has a head shaped like a cow on earth.

(V) Many other types and marvelous forms of animals, fishes, and birds are found in the northern regions, about which a great book could be produced. For as the torrid zone in Africa has its unique and marvelous animals which can scarcely live outside the heat of that region, so in turn the creator gave the cold northern region its own animals which cannot bear the heat of the sun.

Sebastian Münster's Monstra Marina & Terrestria, *a plate in multiple editions of* Cosmographia, *beginning in 1544. It is the second of the three preeminent representations of sea monsters in Renaissance cartography. The northern sea beasts are adapted from the western third of Olaus Magnus's* Carta Marina. *The wolves, wolverine (gulon) between trees, reindeer, snakes, and others in the top panel of the chart are adapted from the terrestrial portions of Olaus's map (they are annotated as E i, B g, B i, and F c, respectively).*

Abraham Ortelius's
ISLANDIA

The third famous keyed chart of marine beasts of the northern seas is a map of Iceland, published by the great Flemish cartographer, Abraham Ortelius (1527–1598). *Islandia* was printed in the 1590 and later editions of Ortelius's *Theatrum Orbis Terrarum*, which is considered the first modern world atlas. Over nearly half a century, the ever-growing collection was published in scores of editions in Latin, English, and the European languages.

In his introduction to the Iceland map, Ortelius often cites Olaus's writings on Scandinavia. However, with its place names, mountains, glaciers, and even an erupting volcano, this Iceland is more accurate and detailed than it is on Olaus's map and others—and most of the monsters, too, have evolved from those rendered by Olaus and Münster.

Evolving Sea Monsters

Two of the *Islandia* figures are adapted copies of those on the *Carta Marina*: the ray gripping a shark (I) and the ziphius swallowing a seal (E). Although Ortelius's engraver might have modeled the Ziphius more directly on Münster's beast than on Olaus's, this creature lacks the wedged back of its predecessors. Above the ziphius, the sea swine (D) differs from the beast that Münster copied from Olaus, but it retains the three eyes on its body. In his key to the "hyena or sea hog," Ortelius cites Olaus and Book 21 of his *History*. With the notable exception of the elegant, decorative hippocampus (G), most of the others are more naturalistic counterparts of monsters that are depicted on the other two charts. Unlike the beasts of Olaus and Münster, they have specific names.

In keeping with *Islandia*'s more objective cartographic purpose, violent scenes of endangered swimmers, ships, and sailors are conspicuously absent. So, too, are the attacks of the orca and "another grisly monster" on fellow creatures. Olaus's fantastic rhinoceros-like hybrid and giant lobsters are nowhere to be seen. *Carta Marina*'s patches of Spermaceti and polar bears on ice floes, however, are present here as well.

Abraham Ortelius's Islandia, *from later editions of* Theatrum Orbis Terrarum *(1570 and after). It is the third of the three preeminent representations of sea monsters in Renaissance cartography. The figures are indebted to those on Olaus Magnus's map and Sebastian Münster's chart. As Joscelyn Godwin notes in his* Athanasius Kircher's Theatre of the World *(2009), the sea monsters were dropped from Johann Jansson's 1612 version of* Islandia: *"The time of maps filled with imaginary denizens of land and sea was passing."*

SEA MONSTER KEY

(A) is a fish, commonly called **NAHVAL**. If anyone eats of this fish, he will die immediately. It has a tooth in the front part of its head standing out seven cubites. Some have sold it as the unicorn's horn. It is thought to be a good antidote and powerful medicine against poison. This monster is forty ells [forearms] in length.

(B) The **ROIDER** is a fish of one hundred and thirty ells in length, which has no teeth. The meat of it is very good, wholesome, and tasty. Its fat is good against many diseases.

(C) The **BURCHVALUR** has a head bigger than its entire body. It has many very strong teeth, of which they make chess pieces. It is 60 cubites [forearms] long.

(D) The **HYENA** or sea hog is a monstrous kind of fish about which you may read in the 21st Book of Olaus Magnus.

(E) **ZIPHIUS**, a horrible sea monster that swallows a black seal in one bite.

(F) The English whale, thirty ells long. It has no teeth, but its tongue is seven ells in length.

(G) **HROSHUALUR**, that is to say as much as sea horse, with manes hanging down from its neck like a horse. It often causes great scare to fishermen.

(H) The largest kind of whale, which seldom shows itself. It is more like a small island than like a fish. It cannot follow or chase smaller fish because of its huge size and the weight of its body, yet it preys on many, which it catches by natural cunning, which it applies to get its food.

(I) **SKAUTUHVALUR**. This fish is fully covered with bristles or bones. It is somewhat like a shark, but infinitely bigger. When it appears, it is like an island, and with its fins it overturns ships.

(K) **SEENAUT**, sea cow of gray color. They sometimes come out of the sea and feed on the land in groups. They have a small bag hanging by their nose with the help of which they live in the water. If it is broken, they live altogether on land, accompanied by other cows.

(L) **STEIPEREIDUR**, a most gentle kind of whale, which for the defense of fishermen fights against other kinds of whales. It is forbidden by proclamation that any man should kill or hurt this kind of whale. It has a length of at least 100 cubits.

(M) **STAUKUL**. The Dutch call it *Springval*. It has been observed to stand for a whole day long upright on its tail. It derives its name from its leaping. It is a very dangerous enemy of seamen and fishermen and greedily goes after human flesh.

(N) **ROSTUNGER** (also called *Rosmar*) is somewhat like a sea calf. It goes to the bottom of the sea on all four of its feet, which are very short. Its skin can hardly be penetrated by any weapon. It sleeps for twelve hours on end, hanging on some rock by its two long teeth. Each of its teeth are at least one ell long and the length of its whole body is fourteen ells.

(O) Spermaceti or a simple kind of amber, commonly called **HUALAMBUR**.

MYTHICAL ANCESTRY

The lineage of Olaus's sea beasts can be traced back to the misty reaches of the human myth pool.

The First Sea Monster

The primeval sea filled the dark abyss long before the sky and earth were named. So declares the Babylonian epic of creation. The mother goddess Tiamat was the salt water of that primordial ocean and her husband, Apsu, the fresh water. Their mingling produced the generations of gods. After a son, Ea (Enki), murdered the father, the vengeful Tiamat spawned a horde of monsters to destroy the usurper: demons and hybrid men that had the bodies of scorpions, bulls, and fish. Ea's son, Marduk, slew her with arrows and from her split body created earth and sky.

Since Tiamat was the sea and is often regarded as a serpent or dragon, she might well be considered the first sea monster, and her fish-man *(kullulû)* one of the earliest mermen. Ea, in turn, established his kingdom in the watery netherworld. Among his retinue of guardian creatures were a fish-man and a goat-fish *(suhurmasû)*, an ancient ancestor of land-sea animals. The Romans called that horned figure Capricornus, and reinterpreted it as a form of Pan. Capricorn is the tenth astrological sign in the Zodiac.

"The Death of Leviathan," a wood engraving after the drawing by Gustave Doré (1832–1883), from Doré's illustrations for the English Bible (1865). The engraving depicts the "piercing serpent" and "crooked serpent" form of the monster whose name is also associated with great whales.

Stone relief of a swimming fish-man (kullulû, *merman) from the first-millennium palace of the Assyrian king Sargon II (ca. 721–705 BC), in modern Khorsabad, Iraq. The figure is related to the fish-man Oannes, typically pictured as a man with a fish's head and scaly robe or as a fish from the waist down.*

The Leviathan

Leviathan, like Tiamat, emerged from the cosmic darkness. Immense and monstrous, this fearsome beast is evoked in different shapes in numerous Judeo-Christian writings. David the Psalmist cites it in a passage on the wonders of creation:

> *. . . this great and wide sea, wherein are things creeping innumerable, both small and great beasts.*
>
> *There go the ships: there is that leviathan, whom thou hast made to play therein.* (Psalms 104:25–26)

That leviathan could easily be a whale, for which there was no word in Hebrew.

However, earlier, in Psalms 74:14, God "brakest the heads of leviathan in pieces" and fed him to the righteous in the wilderness. That fierce beast takes on distinctly reptilian forms in Isaiah 27:1, where God with his sword will "punish leviathan the piercing serpent, even leviathan that crooked serpent." The best-known biblical portrayal of leviathan is in Job 41: "Canst thou draw out leviathan with an hook?" This fire-breathing monster, whose teeth are "terrible round about" and whose scales "are joined one to another," is usually identified with the crocodile.

Leviathan embodies the powers of the sea as its terrestrial counterpart, behemoth (regarded as a hippopotamus), does the land. All leviathan's forms notwithstanding, the name is commonly interchangeable with "whale," which is the largest-known animal ever to live on the earth.

A Nereid on a serpentine, crested ketos. *The lid of a silver* pyxis *(a circular box) found at Canosa di Puglia, Italy, ca. 130 BC.*

A fourth-century mosaic fragment of "The Triumph of Neptune and Amphitrite," from a Roman villa in Algeria. The sea god Neptune (Poseidon) and his wife ride in a chariot pulled by sea horses. Below them, realistic fish and cephalopods (types of marine mollusk) swim amid cupids fishing and riding on dolphins.

Greco-Roman Lore

Homer's "wine-dark sea" is ruled by Poseidon Earth-Shaker, who travels in a horse-drawn chariot. He and his Nereid queen, Amphitrite, produced a brood of sea creatures, one type of which was the *ketos,* Greek for "sea monster or large fish." The word's earliest literary uses are found in the Homeric epics. In the sea, after losing his raft in a storm raised by Poseidon, Odysseus fears "a sea monster, of whom Amphitrite keeps so many." The line is repeated when Odysseus uses *ketos* as a possible food source for the hideous Scylla.

Ketos later becomes a named sea monster when Poseidon sends it to ravage the lands of King Ceteus and his vain queen, Cassiopeia, who boasted she was more beautiful than the Nereids. Ancient Joppa (now Jaffa, Israel) is the traditional site of Perseus slaying the beast to save the sacrificial daughter, Andromeda, and it is a center for sea monster tales. Jonah sailed from there, and in the same area, St. George slew the dragon from the marshes. The renowned *ketos* (*cetus* in Latin), Perseus, and Andromeda myth is told in constellations, in Greek and Roman art, and in paintings up to our own time. "Monstrous sea beasts" are among Neptune's retinue in Virgil's *Aeneid,* and joyous processions of nymphs and Tritons riding hippocampi abound in Roman sculpture, frescoes, and mosaics. These figures are revived as decoration on Renaissance maps while Olaus's sea monsters gradually lose their fearsome power.

NATURAL HISTORY INHERITANCE

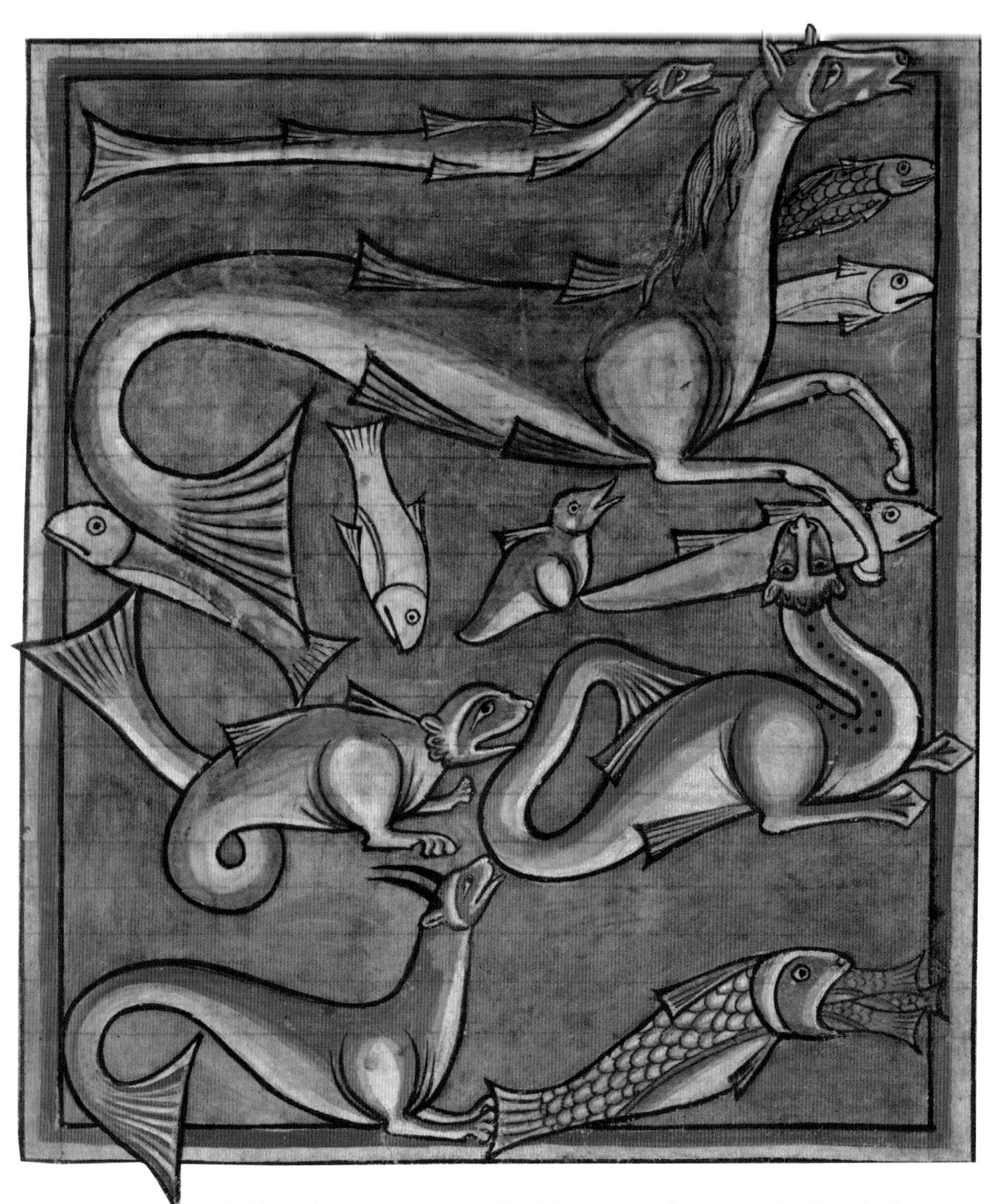

A thirteenth-century bestiary illumination introducing the final section of that book of beasts: fish (pisces). *A sea horse, dragon, and animal-faced creatures mix with more realistic fish. The scribe's text from Isidore of Seville's seventh-century* Etymologies *explains that some fish are amphibians that live on the land and in the sea, and others in the sea are named by their color, shape, or nature that is similar to animals on land. Harley MS 4751, f.68, Cotton.*

As an ecclesiastic and a scholar, Olaus inherited the written knowledge of his time. Most of Europe's pagan mythological stories had already been relegated to the realm of poets, replaced by Judeo-Christian belief. Olaus was free to draw from scriptural, classical, and medieval authority. He combined these with the Scandinavian history and folklore he gathered from his extensive travels. The sea beasts of the *Carta Marina* and his *History* thus emerged from traditional natural history and regional lore, and they influenced not only maps but also the beginnings of marine biology.

Classical Origins

The West's systematic study of animals begins with Aristotle's fourth-century BC *Historia Animalium.* Aristotle cites authorities and contemporary reports, but he depends most on personal observation. He writes that his investigation of animal life can proceed "only after we have before us the ascertained facts about each item." Because his works are accepted as definitive, the use of his scientific method is rare prior to Conrad Gesner's own *Historiae Animalium* in Olaus's time.

Olaus cites Aristotle frequently, but of the hundreds of sources he acknowledges in his *History,* Pliny the Elder (AD 23–79) is the author he draws upon most often. For good reason. Pliny's *Natural History* had a greater impact on medieval natural history than did Aristotle's zoological treatises.

The vast encyclopedia contains little of Pliny's first-hand investigation. The work is a compilation of about 500 authorities from 2,000 consulted volumes. A great number of the selections in *Natural History* are travelers' tales that establish a host of fantastic creatures among the animals of earth.

Religious and Pagan Sources

Directly or indirectly, Pliny's stories intertwine with the pagan and Judeo-Christian sources of Mediterranean animal lore known as the *Physiologus* ("The Naturalist"). This collection of fifty legends, which are part of religious teachings, is thought to have been composed in or near Alexandria as early as the second century AD. As the popular book evolved over time, one of the Church fathers named as a possible contributor was St. Ambrose (ca. AD 339–397), whose *Hexameron* is an important source for Olaus.

Another Church father and Olaus source, Isidore of Seville (ca. 560–636), drew heavily on the writings of Pliny and other authors indebted to him. The animal chapters of his *Etymologies*, an encyclopedia of secular "science," do not contain moralistic allegories; the classical lore of his book expands the *Physiologus* into the illuminated bestiaries of the eleventh to thirteenth centuries. Those celebrated books of beasts, birds, fish, and reptiles often return to the lessons of *Physiologus* and treat both actual and fabulous animals as members of God's animal kingdom.

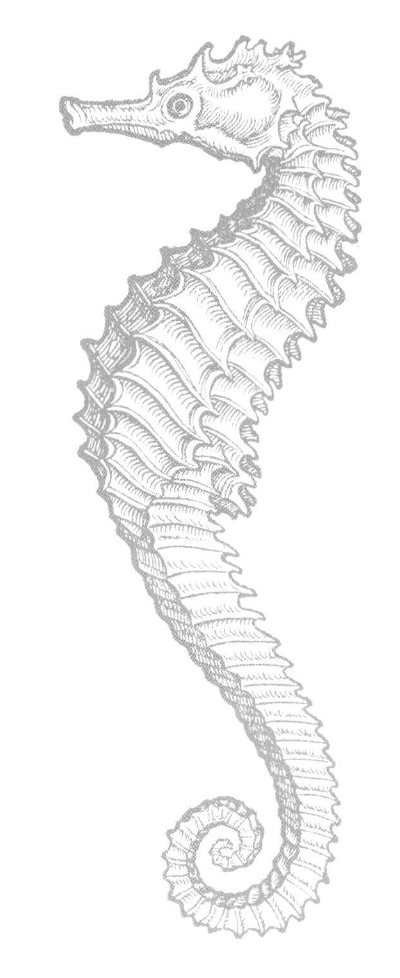

Two sea horses from Conrad Gesner's Historiae Animalium, *which is regarded as the beginning of modern zoology. The monumental volumes set out to collect all that was known about any animal, from traditional lore to modern observation. Gesner represents both the fabulous hippocampus of Neptune (from Pierre Belon's* De Aquatilibus*) and its zoological counterpart, the* Hippocampus guttulatus*, a close relative of the pipefish. Gesner reproduced many of Olaus Magnus's* Carta Marina *sea monsters with the disclaimer that Olaus himself was responsible for their accuracy and that he might have been influenced more by mariners' tales than by actual animals.*

Beginnings of Modern Zoology

The natural history of Albertus Magnus (ca. 1200–1280) is different. His *De Animalibus* is the first systematic natural history since Aristotle, and the only one prior to the sixteenth century. Yet Albertus's book had little immediate influence on the approach to natural history. Medieval encyclopedias of Vincent of Beauvais (?–died 1264), Thomas of Cantimpré (1210–1275?), and others continued to transmit animal lore from Pliny and other traditional authorities. But while bestiaries fade in importance during the Reformation, naturalists begin to return to Aristotle's scientific method.

Between the printing of Olaus's 1539 *Carta Marina* and the 1555 *History,* modern zoology was being born, and with it the observation and classification of marine life. Seminal works of ichthyology by Pierre Belon (1517–1564) and Guillaume Rondelet (1507–1566) appeared during the short span of years that saw the publication of the first three volumes (1551–1555) of Conrad Gesner's *Historiae Animalium.* His fourth volume, on fishes (1558), reproduced much work from both Belon and Rondelet—and from Olaus's map and book. As scientific as all these transitional studies attempt to be, they inevitably contain some woodcuts that are derived more from the imagination than from observation of real life.

The voyage on the *Carta Marina* takes us up a seam in time.

"There are monstrous fish on the
Coasts or Sea of Norway"

OLAUS MAGNUS

VRSI ALBI
MARE GLACIAE
DANI
L
XII·P
ANGLI
L
GOTHI
OOPEDVM

THE VAST OCEAN

At the outset of a voyage up the northern seas of the *Carta Marina*, Olaus Magnus speaks of the expanse and mystery of the waters of the world. His paean to the sea and its counterparts to land and sky relies heavily on ancient authority.

> *"The vast Ocean in its Gulph offers all Nations an admirable spectacle, and shews divers sorts of Fish; and these not onely wonderful for magnitude, as the Stars are compared one with another, as they are terrible in shape; so that there is nothing in the Ayr nor Earth, nor Bowels of it, or in domestique Instruments that may seem to lye hid, that is not found in the depth of the Sea. For in the Ocean that is so broad and by an easie and fruitful increase receives the Seeds of Generation, there are found many monstrous things in sublime Nature, that is always producing something; which being perplexed and rolled up and down one upon another by the ebbing and flowing of the Waters, they seem to generate forms from themselves and from other principles; that whatsoever is bred in any part of nature, we are perswaded is in the Sea, and many things are to be found there that are to be found nowhere else.*
>
> *And not onely may we understand by sight that there are Images of Animals in the Sea, but a Pitcher, a Sword; Saws and Horses heads apparent in small Shell fish. Moreover you shall find Sponges, Nettles, Stars, Fairies, Kites, Monkies, Cows, Woolves, Mice, Sparrows, Black-Birds, Crows, Frogs, Hogs, Oxen, Rams, Horses, Asses, Dogs, Locusts, Calves, Trees, Wheels, Beetles, Lions, Eagles, Dragons, Swallows and such like: Amongst which, some huge Monsters go on Land and eat the roots of Trees and Plants: Some grow fat with the South wind; some with a North wind blowing.*
>
> *Also, I must add, that on the Coasts of Norway, most frequently both old and new Monsters are seen, chiefly by reason of the inscrutable depth of the Waters. Moreover, in the deep Sea, there are many kinds of fishes, that seldome or never are seen by men."*

CHARTED COURSE ON THE *CARTA MARINA*

— H —

(F) The Rockas and sharks

— E —

(M) The Sea Worm and giant lobster

— D —

(I) The Duck Tree

(M) The Polypus ("Octopus")

(L) Balena and Orca

(K) The Sea Swine

— A —

(1) Southwest, not keyed, The Sea Unicorn

— D —

(O) The Prister

(D) The Ziphius

(E) "Another grisly monster"

— E —

(3) West, not keyed, The Sea Cow

(E) A Rhinoceros-like Monster and giant lobster

— D —

(P) Spermaceti

(A) Beached Whale

— A —

(K) More Pristers

(L) The Island Whale

— B —

(D) The Sea Serpent

(F) Caribdis

(C) Another Prister

(2) Northwest, not keyed, A Sea Creature

(B) A Rosmarus

(B) The Kraken (Giant Squid)

This book's charted voyage begins at the bottom of the map and winds northward through the vast ocean's Nordic seas, in whose waters Olaus's sea beasts endanger fishermen and sailors. Like modern whale-watching tours, this course is set to sail in the map's areas where certain beasts are known to be.

A
B
CARTA MARINA ET DESCRIPTIO SEPTEMTRIONALIVM ERRARVM AC MIRABILIVM RERVM IN EIS CONTENTARVM
GRVTLADIE PARS
GRVTLANDIE PARS
ISLANDIA
FINMARCH
LAPPIA
BOTNIA OCCIDENTALIS
FARE
ZIPHIVS
TILE
BALENA
SCANDIA
NORVEGIA REGNVM
SVECIA
DALACARLIA MAGNA
VERMIS·40·PE·LO
MARE DEVCALIDONICVM
SCOCIE PARS
OCEANVS BRITANICVS
ANGLIE PARS
MARE GERMANI
DANIA
GOTHIA
MARE GOTICVM
D
E
G
H

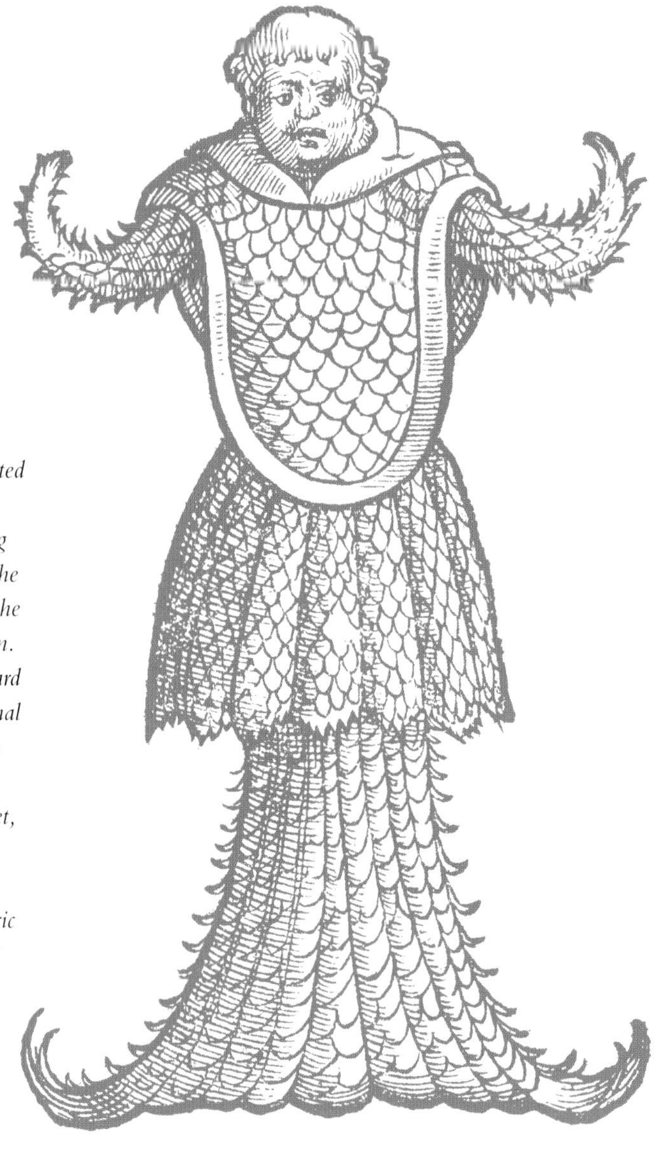

A sea monster "like a Fryer with his Cowl." Conrad Gesner's Sea Monk, (or Monk-fish), reproduced from Guillaume Rondelet's book of fishes and adapted in many Renaissance natural histories. During Gesner's own lifetime, the monster was caught in the Baltic, near Copenhagen. Marine authority Richard Ellis concludes the animal was a giant squid (The Quest for the Giant Squid, 1998). Rondelet, Gesner, and others pair the Sea Monk with the Sea Bishop, another cleric figure, based on a 1531 sighting near Poland.

Dangerous Fishing

Olaus continues, turning from ancient authority to what he has observed and what he has gleaned from fishermen who risk their lives on the Norwegian Sea:

"The Fishing is said to be dangerous in the Norway Ocean for many Reasons, because men fish in the open Sea very far from Land: When great Tempests arise, the Fisher-men are soon drown'd by the Waves: Where great Sholes of Ice flote, they are dispersed: by the fishing of Whales, and other Monsters they are distracted. Moreover, sometimes they are debilitated in their hands by some monstrous Fish out of the Sea; and if they do not presently let them go a Tempest riseth and drowns them. If therefore some rash Fisher-men, fighting with some Sea Monster, pull him into the ship, that is like a Fryer with his Cowl, they are presently overwhelmed with howling and crying of these Monsters, that they can neither cast forth their Hooks to catch fish nor row with their Oars, and they can scarce hoise [hoist] *up Sail to be gone, unless they let go the Monster.*

Many Thousands of Fisher-men dwell in the Villages in the utmost Borders of Norway. In February and March, and also in January, the Inhabitants of this Country go in strong ships to fish from the shore into the Deep: as far as they can sail in two days, carrying with them necessaries for their Food for 20 or 30 days. But the place where they most frequently exercise their Fishing is between Norway and Island. Nor do the Fisher-men, when it is Tempestuous Weather, ride at Anchor, but they fish floting up and down till their ships be full. And it is observed that when ever monstrous fish are drawn forth of the Sea; with men or Lions faces, and the like, that this always foreshews Discords and War in the Land."

A vignette from Olaus Magnus's History chapter. Olaus Magnus scholar John Granlund notes the fish are cod and a halibut and that ice-fishing was not done in the north Norway waters.

Northern Ships

Fishermen and merchants ply the waters of the *Carta Marina* in crafts of many sizes and kinds: from rowboats, rafts, and leather boats of the Greenlanders to sailing ships of the northern European nations. While ships that decorate the open spaces of maps are often generic, Olaus's are not. Like the tales he heard from Scandinavian fishermen, his boats and ships derive from his first-hand experience of living in the Baltic port city of Danzig and traveling throughout Scandinavia.

The medieval merchant vessels in the North Sea were influenced by the ships of the Norsemen. Like their predecessors, Hanse cogs—cargo boats of the thirteenth and fourteenth centuries—were also "clinker-built" with overlapping hull planks, with a single square sail. Their side rudder, however, was moved to the stern, under a fighting platform. These were the ships of the Hanseatic League, a federation established by northern German cities to protect their trading interests. While the alliance's centuries long dominance of Northern European trade routes was fading, the three-masted carracks of southern Europe transformed shipbuilding in the North. These were typically "carvel-built," with joined hull planking.

Carta Marina scholar Edward Lyman points out that the model for most of Olaus's larger ships would seem to be a Flemish form of the armed carrack. The standard carrack had three masts and raised castles fore and aft.

The three-masted Danish carrack under full sail. From the southwest corner of the Carta Marina, *the starting point of the imaginary voyage up the map.*

Northern ships in Pieter Bruegel the Elder's Landscape With the Fall of Icarus *(ca. 1558). The mythological event is pictured as the hubristic boy's legs disappear into the water, passed by the closest ship and unnoticed on shore.*

Unlike the square-rigged fore- and main-masts, the mizzenmast was lateen-rigged with a triangular sail like the Portuguese caravels. The foremasts of Olaus's ships are strangely tilted forward.

The chart's legends identify ships from Denmark, Holland, Norway, England, Scotland, Hamburg, Lübeck, and Bremen (the latter three in the Hanseatic League). These international sea-lanes are crowded and potentially contentious. In the far north, the Hanseatic Hamburg ship is sinking the Scottish vessel with cannon fire as the two vie for trade in Iceland. The most elegant ship—with pennons flying—is the Danish vessel at the bottom of the map. That is the point at which this book's voyage begins.

THE ROCKAS

The anonymous poet of the Old English "The Seafarer" said it: There is a time for journeys, when your spirit sends you beyond the shore to the whales' road. He also said there is no one on earth who knows no fear when the sail unfurls. A journey up the *Carta Marina* soon proves the poet's wisdom when the first sea creatures surface off the southwest coast of Norway. Olaus's commentary derives from the *History* chapter, "Of the cruelty of some Fish and the kindness of others."

"There is a fish of the kind of Sea-Dog-fish, called Boloma, *in Italian, and in Norway,* Haafisck, *that will set upon a man swimming in the Salt-Waters, so greedily, in Troops, unawares, that he will sink a man to the bottome, not onely by his biting but also by his weight; and he will eat his more tender parts, as the Nostrils, Fingers, Privities, until such time as the Ray come to revenge these injuries, which runs thorow the Waters armed with her natural fins & with some violence drives away these fish that set upon the drown'd* [soaked] *man, and doth what he can to urge him to swim out. And he also keeps the man, until such time as his spirit being quite gone, and after some days, as the Sea naturally purgeth it self, he is cast up. This miserable spectacle is seen on the Coasts of Norway, when men to wash themselves, namely strangers and Marriners, that are ignorant of the dangers, leap out of their ships into the Sea. For these Dog-fish, or* Boloma, *lie hid under the ships riding at Anchor, as Water-Rams, that they may catch men, their malicious Natures stirring them to it. But the Urinators* [divers] *avoid this danger with sharp stiles* [shafts] *tyed by lines; for with these, they kill these Sea-Rams, and Dog-fish: and unlesse they be stricken through with these, they will scarce retreat. So cruel a fight is fought with them under water … The Skin of this Dog-fish, for the roughnesse of it to polish wood and bones, is of the same nature with a Rays Skin."*

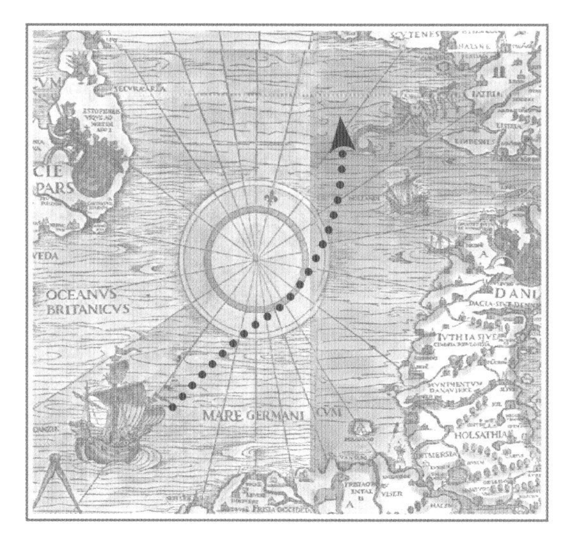

The voyage begins in the German Sea off the kingdom of Frisia, known for its fine horses. It progresses northeast past Denmark and a Dutch ship, and bears northerly along the coast to the first sighting of sea creatures.

SVINBORHOVIT
M
RD
QVIVEFIO
TILEMARCH
VALDRES
EMPORIVM
MAXIMVM
BERGE
HVITINGO
LERDALHADD
HARDANGER
VERMIS·40·PE·LŌ
SKERE
BAGGA
HOLM
KRABBA
KIRXIE
SVLLOPP
NIDAL
SOL
MARK
SNLODAL
NOT
E
HIELMELAND
SVDERH
XII
VILLE
SCVTENES
CARMESVND
FESTIA
NOTE
COPERVIK
LIDER
IATRIA
TONSBERG
SKIDENI
AGDASIDA
LISTRIA
F
LINDESNES
DOM REGI
HESIME
FLEDERE IŌFRV·SVD
LONGA
HOLLANDI
BOVEBER
HOLM
OPORTET·IVSTV
ET HVMANV ESSE
SAP·12
RICOPE
A
VIBVRG
DANIA

THE ROCKAS

Carta Marina

According to the explanatory *Carta Marina* key (H f), a benevolent fish called Rockas attempts to save a swimmer "from being devoured by the sea monsters." The generic term, "sea monsters," generally refers to enormous marine animals that were regarded as dangerous. These tend to be mammals rather than fish, but they can, of course, be various threatening creatures, with different regional names or no specific name at all. These "monsters" are small sharks, members of the prehistoric ocean family feared up to our own time. The ray on the *Carta Marina* that comes to the swimmer's rescue is even more "monstrous" in size than the ferocious "Sea-Dog-fish," but Olaus casts it as friendly to man.

Ancestral Lore

Many of the details in Olaus's story come from his favorite classical source: Pliny's *Natural History* (ca. AD 77–79). Nonetheless, Olaus's version of the benevolent ray seems to be his own. It grows out of a long folklore tradition best known through Aesop's fable of the slave Androcles, who befriends a lion that later saves him from a gladiator's death.

Elsewhere in his *History* chapter, Olaus compares his ray's attempts to nurse the swimmer back to health to Albertus Magnus's tale of dolphins carrying a live or even dead swimmer to shore on their backs—unless the dolphins could smell that the unfortunate man had eaten dolphin flesh. In that case, they would devour him. In the same entry of his natural history, Albertus recounts classical dolphin stories that illustrate the tradition shared by Olaus's kindly ray. One is that of the "boy on a dolphin," whose love for the animal was so great and so mutual that after the boy's death, the dolphin dies of loneliness. Also well known is his story of Arion, a young harper. Threatened on a sea voyage, he plays for the sailors before leaping into the sea and being saved by dolphins drawn to his music.

Olaus later cites another sea beast that rescues a seafarer tossed overboard by a crew. The whale swallowed Jonah "to grant him life." So wrote St. Ambrose (ca. AD 339–397), whom Olaus quotes, and he compares Jonah's restoration with the Resurrection of Christ.

A third-century AD Roman mosaic from Thaenae baths, Tunisia, North Africa. The image depicts the legendary Arion playing his cithera on a dolphin that rescued him after he escaped from a hostile crew by leaping overboard. Earlier in his History*, Olaus tells of seeing dolphins being drawn to a harpist's music out on the open sea. After they churned the waters in appreciation and swam off, a fierce storm arose.*

Map Legacy

The Rockas clearly indicates the influence of Olaus's sea beasts on the charts of Sebastian Münster and Abraham Ortelius.

Münster's image of the ray with a large fish in its mouth and a single fish attacking the swimmer varies Olaus's art only slightly.

The cosmographer's ray key (S) echoes Olaus's, down to the use of the generic "sea monsters" in reference to sharks.

Ortelius's ray (I), without a swimmer, is more naturalistic than either of the other two images and is also the most artistic. His ray key, on the other hand, is not at all similar to the other keys. The beast has a Dutch name, and the text strangely juxtaposes accurate natural history details with Island Beast lore: "SKAUTUHVALUR. This fish is completely covered with bristles or bones. It is somewhat like a shark or skate, but infinitely bigger. When it appears, it is like an island, and with its fins it overturns boats and ships."

Ortelius correctly links the shark and skate, which share cartilaginous—as opposed to bony—vertebrae. The rays and skates of the northern seas, however, are considerably smaller than the legendary manta ray of southern waters.

The related ray images of the three sea monster charts comprise a cluster all their own, with little evident influence on natural history illustrations.

And Since

Olaus notes in his *History* that both rays and dogfish can be eaten. Only three years after publication of Olaus's book, naturalist Conrad Gesner revealed another use of rays in the fourth volume of his *Historiae Animalium* (1558). Beside a woodcut of a monstrous reptilian "basilisk" exhibited in Venice, he explains that, "The apothecaries and vagabonds sculpt the bodies of the rays in various forms of their own pleasing by cutting/bending/splitting into the forms of Snakes/Basilisks and Dragons." Such hoax specimens became popularly known as "Jenny Hanivers." Despite Gesner's debunking of such fakes, Ambroise Paré (1510–1590) later presents one as a flying fish in his *Of Monsters and Prodigies.*

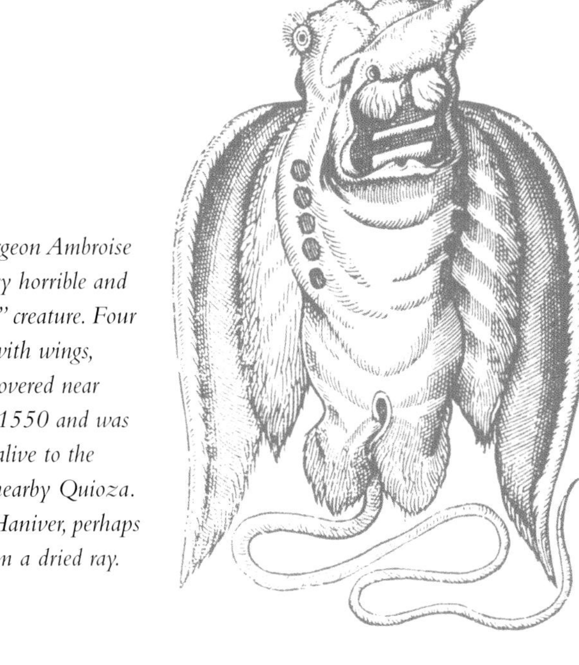

French surgeon Ambroise Paré's "very horrible and monstrous" creature. Four feet long, with wings, it was discovered near Venice in 1550 and was presented alive to the mayor of nearby Quioza. A Jenny Haniver, perhaps created from a dried ray.

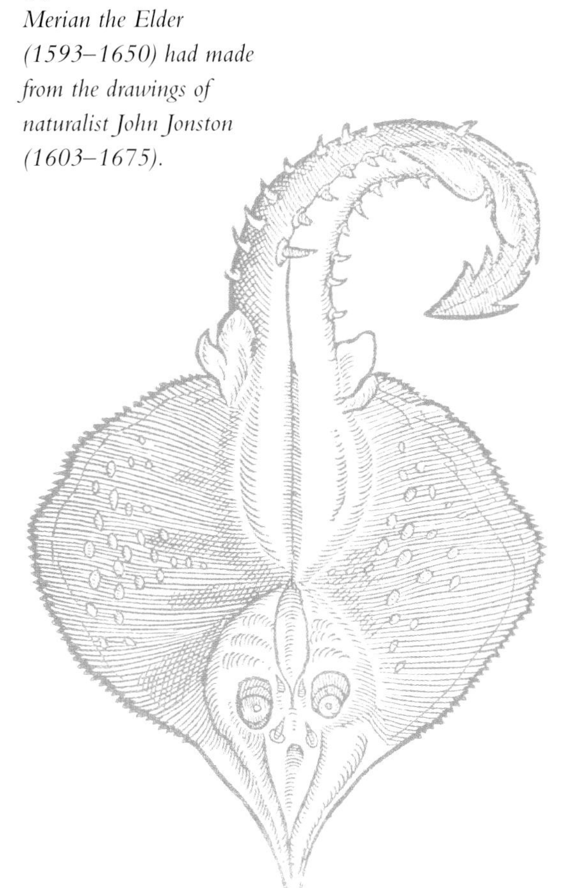

A seventeenth-century engraving of a thornback ray (Raja clavata) *that resembles the* Carta Marina*'s Rockas, from* Theatrum universale omnium Animalium *(after 1650), a collection of plates that Matthäus Merian the Elder (1593–1650) had made from the drawings of naturalist John Jonston (1603–1675).*

In our time, relatives of both kinds of sea beasts in Olaus's account of the cruelty and kindness of marine animals have been the basis of best-selling novels by Peter Benchley. The Great White Shark of his celebrated novel, *Jaws* (1974), and its record-breaking screen adaptation by Steven Spielberg, unleashed a global wave of fear of going near the water. The manta ray in *The Girl of the Sea of Cortez* (1982) is another matter. Rescued from a fishing net by a girl diver, the ray later saves the girl's life in a story that follows the tradition of Androcles's lion and Arion's dolphins.

THE SEA WORM

Map E (M)

The voyage proceeds up the coast to a sighting of what the *Carta Marina*'s legend identifies as a *vermis*. The key describes the writhing, delicate creature as, "A sea snake, 30 or 40 feet long." Olaus's commentary derives from a chapter entitled, "Of very long Worms." He enhances the account with his personal experience and with lore he gathered from his years of travel and talking to Nordic mariners.

"There is, on the Coasts of Norway, a Worm of a blew and gray color, that is above 40 Cubits long, yet is hardly so thick as the arm of a child. He goes forward in the Sea like a Line, that he can hardly be perceived how he goes; he hurts no man, unless he be crushed in a mans hand: for by the touch of his most tender Skin, the fingers of one that toucheth him will swell. When he is vexed and tormented by Crabs, he twines himself about hoping to get away, but cannot. For the Crab with his Claws, as with Toothed Pincers, takes so fast hold at him, that he is held as fast as a ship is by an Anchor. I oft saw this Worm, but touched it not, being fore-warned by the Marriners."

The *History* vignette in which two sea worms appear along with a blossom and stars with human faces, and creatures with feet and ears, is unusually fanciful. It illustrates adjoining chapters that the worm shares with sponges, sea nettles (sea anemones), and starfish. The sponge feeds on little fish, and the nettles irritate whatever they touch. Traditionally, they are neither plants nor animals, but combine the natures of both. Starfish are the sea's counterpart of the stars of the heavens. They are said to contain such great heat that anything they devour is cooked inside them.

Olaus varies the length of the worm when described in the map's key and the *History* text. While the key stipulates that the sea snake is up to 40 feet in length, it is 40 cubits (60 feet) in Olaus's book.

The Sea Worm is being attacked by a giant crab at the southern border of the Kingdom of Norway, where the king sits upon a throne. In the shield beside him are the arms of the kingdom: a lion gripping a broadax.

SOGNIA
NEMO·ACCIPIAT
CORONĀ·TVAM
APŌ·3
STAD
HERŌ
OSTER
VOOS
VALDRES
ASTRE
HALSNE
KLYFTA
MVI
BRYSTSV ND
SKALVOGH
NORVEGIA
REGNVM
HALIMGALAB
TOTEN
SOLO GIER
SVNDARDAL
QVIVEFIO
NBORHOVIT
M
TILEMARCHIA
EMPORIVM MAXIMVM
BERGĒ
VALDRES
LERDALHADD ADA
HVITINGO
HARDANGER
SVLLOPPAMONS
VERMIS·40·PE·LŌ
SKERĒ
BAGGA HOLM
KRABBA KIRKIE
NIDAL
SOLOGIA
MARK
EDSDAL
SNLODAL
NOTE
HIELMELAND
SVDERHE
SCVTENES
CARMESVND
FESTIA
NOTE
AKERSHV
COPERVIK
LIDER
IATRIA
TONSBERG
SKIDENI
AGDASIDĀ
LISTRIA
LINDESNES
DOM REGIS
HESIME
F

THE SEA WORM

Carta Marina

Olaus's "very long Worm" is a thin, gentle relative of the great sea serpent that terrorizes sailors farther up the *Carta Marina*'s northern seas. So affirmed Conrad Gesner, the sixteenth-century naturalist now generally regarded as the Father of Modern Zoology, and Edward Topsell (1572–1625), the clergyman who adapted Gesner's pioneering works into English a century later.

Map Legacy

The *Carta Marina* sea snake and crab do not appear on the sea monster charts of Münster and Ortelius, nor on other maps. Cartography is one of the two major areas in which versions of *Carta Marina* sea beast images can be found. The other area is natural history.

Olaus's monsters appear often in Conrad Gesner's monumental *Historiae Animalium*. Gesner reproduced virtual copies of many marine figures from the *Carta Marina*—not from vignettes in Olaus's *History*—in his fourth *Animalium* volume, on fishes (1558, one year after Olaus's death). One of those images is based on Olaus's sea worm. It follows entries on eels and appears along with the woodcut of Olaus's most famous sea monster, the sea serpent, on the same page.

Edward Topsell copied the two Gesner woodcuts, one above the other, in his "Of the Sea-Serpents" chapter in his *History of Serpents*. That chapter, like Gesner's, included text on eels. Topsell's description of the smaller serpent, Olaus's Sea Worm, echoes Gesner's:

A colored German version of Conrad Gesner's sea worms from the fourth volume of his Historiae Animalium. *Gesner attributes the pictures of both of the worms to Olaus. The dramatic power of the larger, more terrifying figure below it will dominate the many other images Gesner reproduces from the* Carta Marina.

Von der Wallschlangen.

Nimm wahr zwo grosse Schlangen schwammen
Von Tenedo über meer herkamen/
Mit viel vmbbringen/so ichs sag/
Hab ich ein grossen grawen drab/
Ihr brust erhuben sie entbor/
Hoch über das wasser hervor/
Ihr har war vornen blutfarb gantz/
So zoch mit jhm das meer den schwantz/ rc.

Von den gelben Meerschlangen.

OLaus Magnus setzt einer gelben Meerschlangẽ gestalt in seine taffel/vnd sagt man finde sie im Balthischen Meer auff dreissig oder viertzig schuchlang/sie schedige niemands/wo sie nit zuvor in zorn gereitzt worden.

Von der Wallschlangen.

BEy Nortwegen im stillen Meer erzeigen sich Meerschlangen zwey oder drey hundert schuch lang/sehr auffsetzig vñ verhaßt den Schiffleüten/also das sie auch zuzeyten ein mann auß dem Schiff hinnemmen/sollen sich vm̃ grosse Schiff schlahen vnd dieselben zu grund richten/sie erheben offt solche krümb über das Meer/das darunder ein Schiff ring durch fahren möchte.

Ein Beyerischer Mönch der Schiltberger genennt/schreibt in seiner Reißhistory/das zu der zeit als er in der Türckey gefangen gelegen/vnzalbar viel jrrdische vnnd Meerschlangen ein Statt im Königreich Genik vmbgeben/die jrrdischen seyen auß den wälden/dero dieselbig glegenheit voll/herfür kommen. Die anderen haben sich auß dem Meer dahin verfügt vnnd neün tag lang gesammlet/also daß sich auß forcht vnd schrecken niemand aussert der Statt begeben dörffen. Dieweil sie aber weder leüt noch veich beleidigten/ließ der Fürst deß Lands gebieten/daß sie niemand schedigen solte/dann solch wunderwerck wurde ohne zweyffel etwas grosses bedeüten/den zehenden tag fiengen sie an mit einander zustreiten/vnnd liessen nicht ab/biß sie die nacht begundt zutrennen/do der Fürst solches vermerckt/reitt er mit wenig reütern auß der Statt den streit zubesehen/zuletzt wichen die Meerschlangen ab. Deß morndrigen tags

There is also in the Swervian Ocean or Balthick sea, Serpents of thirty or forty foot in length, whose picture is thus described, as it was taken by Olaus Magnus, and he further writeth, that these do never harm any man until they be provoked.

Only two decades after Gesner's death, an amateur naturalist, Adriaen Coenen (1514–1587), included watercolors and descriptions of both sea beasts in his personal manuscript on whales and other marine life (1585). The Dutch fish wholesaler and beachcomber adapted Olaus's Sea Worm from the vignette in the *History* and Gesner's sea serpent from the *Carta Marina*. The two serpents appear on different pages of his album.

Lobsters as well as crabs gnaw on the thin, winding body of one of the sea worms. Coenen writes that, "Olaus describes it in his 21st book," and goes on to paraphrase Olaus's description of the animal. He identifies the worm as an eel and illustrates an engraving that made the Renaissance rounds from Pierre Belon's 1553 book of fishes to Gesner's natural history and later to Edward Topsell's book of serpents.

Coenen's "long and very slender eel *(zeeworm)*" is only one of many of Olaus's sea beasts that he describes and/or illustrates in his manuscript, usually with an attribution to Olaus himself. Olaus's images in Coenen's whale book usually derive from vignettes in the *History* or from Gesner's woodcuts. They do not come directly from the *Carta Marina.*

Unlike Gesner's lasting contribution to Olaus's sea monster legacy, Coenen's extensive use of the figures and their lore had little if any influence in his own time. His manuscripts were eventually collected in the Royal Zoological Society library, in Antwerp. Extensive reproductions of his pages, with the text translated from the Dutch, were published for the first time in the 2003 *The Whale Book* (edited by Florike Egmond and Peter Mason, with commentaries by Kees Lankester).

Ancestral Lore

The Gesner/Topsell and Coenen natural histories clearly show that Olaus's "sea snake" grows out of the eel tradition.

One of Olaus's most often consulted authorities, Isidore of Seville (ca. 560–636), compiles eel lore in his *Etymologies*: He writes that the animal's name, *anguilla*, is similar to that of a serpent, *anguis*. Born out of mud, eels are so slippery one cannot hold them. They are said to grow to 30 feet long in the Ganges River in the East. And anyone who drinks wine in which an eel has died will find it so distasteful that he will never drink wine again.

Aristotle had declared that eels were spontaneously generated from mud. Pliny calls them "[l]ong slippery fish," which move in the sea the way land snakes do, "propelling themselves by twisting their bodies." In his popular and entertaining collection of animal lore, Aelian (ca. AD 170–235) tells the tale of a Roman general who adorned his pet moray eel with earrings and a jeweled necklace. The eel would swim to him at his call and was mourned at its death.

Eel tradition notwithstanding, modern *Carta Marina* scholar John Granlund notes that ribbon worms off the coast of Norway can extend themselves to more than 15 feet. Other flat ribbon worms of the North Sea can stretch to many times that length. The ribbonfish, too, is flat but hardly as long, and its red dorsal fin in no way matches Olaus's description of the Sea Worm. The ribbonfish's close relative, however, is the oarfish, which grows to a length of 26 feet and was sometimes thought to be the actual creature that witnesses called a sea serpent.

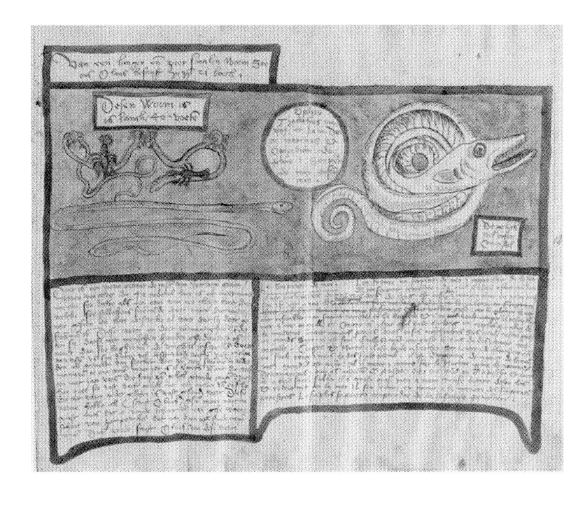

Watercolor eels from the pages of a sixteenth-century personal manuscript that Dutch fish dealer Adriaen Coenen compiled during much of his life. He drew heavily from images and texts of Conrad Gesner, Olaus, Guillaume Rondelet, Pierre Belon, and other naturalists. The worms he attributes to Olaus are adapted from the History *vignette that opens this chapter.*

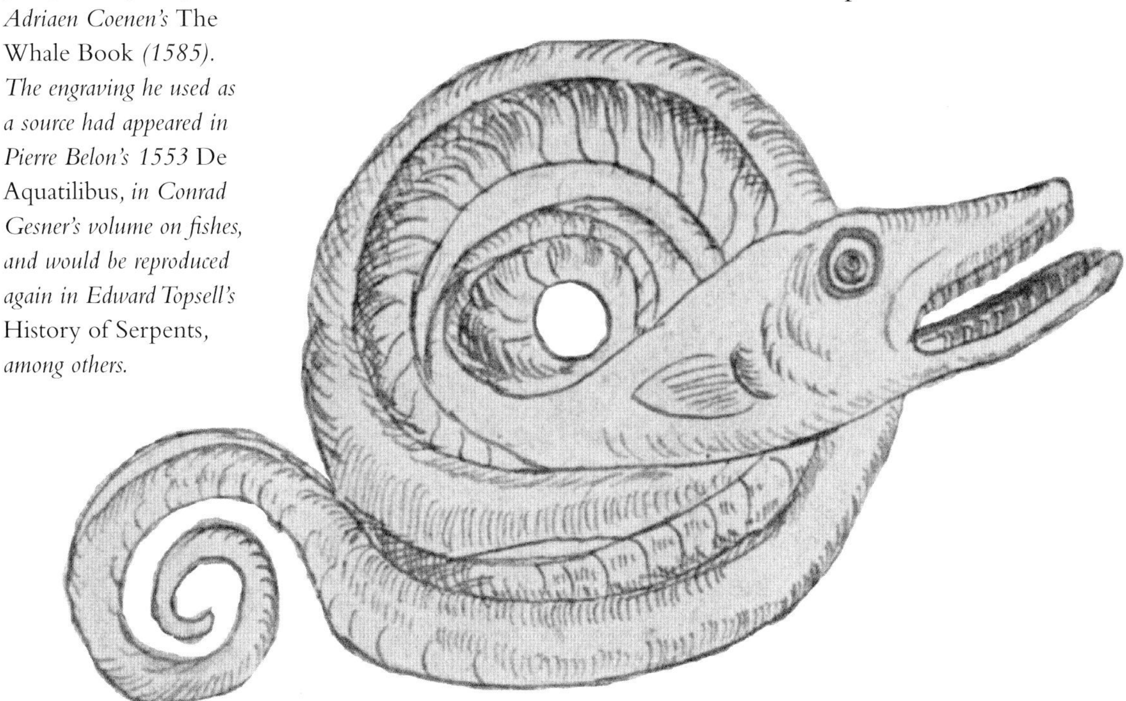

A detail of what is perhaps a conger eel from Adriaen Coenen's The Whale Book *(1585). The engraving he used as a source had appeared in Pierre Belon's 1553* De Aquatilibus, *in Conrad Gesner's volume on fishes, and would be reproduced again in Edward Topsell's* History of Serpents, *among others.*

THE DUCK TREE

Map D (7)

After encounters with violent sea beasts, the voyage approaches the quiet waters of the Orcades (Orkneys) and what the map's key describes as, "Ducks being hatched from the fruit of the trees." The related *History* chapter is "Of the admirable Generation of the Ducks of Scotland." Olaus emphasizes the kind of local natural history that suffuses the land portions of the total *Carta Marina*. Only at the end of his commentary does he return to the enduring fable announced in the *History*'s chapter title.

"[N]ear Glegorn, as a later Scottish Writer testifies, two leagues off is the Rock Bassensis, wherein there is an impregnable Fort, about which there is a strange multitude of great Ducks, which they call Sollendae, which live on Fish: and these are not the same with wild or tame Ducks, in the Species Specialissima; but because they are like them in color and form, they are also called Ducks: but for difference sake Sollend Ducks: These Ducks come yearly by flocks in the Spring, from the South, to the Rock Bassense ... At the end of Autumn, they fly three days about the Rock, and then they fly by Troops to the South parts, to live all the winter, that they may return in Summer; because when it is Winter with us, it is Summer time to those that live in the South ... Moreover, another Scotch Historian, who diligently sets down the secret of things, saith that in the Orcades, Ducks breed of a certain Fruit falling into the Sea; and these shortly after get Wings, and fly to the tame or wild Ducks."

One medieval explanation of where birds migrated to in the winter was they went to the moon. Olaus clearly explains the migration patterns of the "Sollend Ducks," which are known to breed in the South and fly to the cooler North in summer. The Duck Tree of the Orcades solved the mystery of the winter arrival of adult birds that had not been seen mating, building nests, or laying eggs.

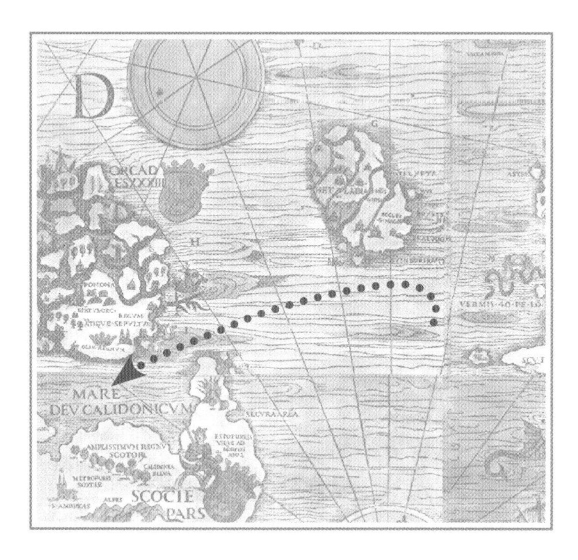

A change of course passes south of the Shetland Islands and proceeds easterly to the Orcades, its shore trees heavy with fruit and its waters dotted with swimming ducks.

ORCAD
ESXXXIII
HET
H
FLE
DERO
POMONA
EPATVS·ORC·
REGVM
ATIQVE·SEPVLTVRE
I
OLIM·REGNVM
SKI
LIVS
MARE
DEVCALIDONICVM
SECVRA·AREA
RMANA
ESTO FIDELIS
VSQVE AD
MORTEM
APO 2
AMPLISSIMVM REGNV
SCOTOR
CALIDONIA
SILVA
METROPOLIS
SCOTIE
SCOCIE
ALPES
·S·ANDREAS
PARS

Olaus's description of ducks being born from trees and the sea is brief, but the *Carta Marina* key and *History* chapter title acknowledge a widespread medieval belief. Accounts of the miraculous birth of marine fowl were already centuries old by Olaus's time—and despite the objections of some scholars, they would persist for another 200 years. Olaus's Duck Tree contributes to one of the most famous learned controversies of the age.

Ancestral Lore

Tales of the Orcades ducks intertwine with earlier stories of the barnacle goose (Bernake, Bernacae) and then the tree goose. No classical writers mention the unnatural births of these birds. It is only in the Middle Ages and the Renaissance that the stories evolve.

Credit for the first written account of the barnacle goose goes to the Welsh ecclesiastic Giraldus Cambrensis (1146–1223). *The Topography of Ireland* contains his personal report of the generation of what he says are Irish birds called barnacles, which are similar to marsh geese but smaller. He writes that, "I have often seen with my own eyes more than a thousand minute embryos of birds of this species on the seashore, hanging from one piece of timber, covered with shells and already formed." From "gummy excrescences" encased in shells, they grow feathers over time and either drop into the water or fly away.

Giraldus's eyewitness details notwithstanding, the legend of a birdlike creature not born in a nest was already known. He chastises Irish clerics for their Lenten practice of feasting on what they regarded as neither flesh nor fowl, but food of the sea. His admonishment might have contributed to Pope Innocent III's issuing of a 1212 edict that prohibited the eating of barnacle goose during Lent. Around that time, Giraldus's manuscript account of the barnacles appeared without attribution in medieval bestiaries. *The Topography of Ireland* was not printed until centuries after, in 1604.

A richly colored thirteenth-century bestiary illumination of geese hatching from trees. The scribe's text excerpts Giraldus Cambrensis's account of the miraculous birth in his Topography of Ireland. *Harley MS 4751, f.36.*

The birth of tree-geese, from the Cosmographia *of Sebastian Münster. The woodcut, with accompanying text, is in the same voluminous work as the* Monstra Marina & Terrestria *plate of figures adapted from Olaus's* Carta Marina.

Not everyone was totally convinced by the miraculous birth stories. Albertus Magnus (ca. 1200–1280) disputed claims that no one had ever seen barnacle geese ("barliates," "boumgans") or tree geese mating or nesting. He claims that "this is absolutely absurd, because I myself and many of my colleagues have observed these birds copulating, laying eggs, and feeding their young."

The spurious John Mandeville, however, was attracted to the marvelous tale of the tree goose. In his fourteenth-century fictional *Travels*, he lifts the legend of the European tree goose from the *Journal* of Franciscan

Odoric of Pordenone (1286–1331). He boasts that, "in our country were trees that bear a fruit that become birds flying."

Map Legacy

Sebastian Münster includes Olaus's Duck Tree in his sea monster chart and varies the *Carta Marina* key only slightly in his own key (I). Earlier in the *Cosmographia*, he writes about the "tree-goose" and includes a detailed woodcut of their birth:

> *In Scotland there are trees which produce fruit conglomerated of their leaves; and this fruit, when in due time it falls into the water beneath it, is endowed with new life, and is converted into a living bird, which they call the "tree-goose." Several old cosmographers, especially Saxo Grammaticus, mention the tree, and it must not be regarded as fictitious, as some new writers suppose.*

He adds a story about the skeptical Aeneas Sylvius (later Pope Pius II,1405–1464), who wanted to see such a tree when he visited Scotland around 1430. After being told it grew farther north, in the Orcades, he complained that, "miracles flee farther and farther."

And Since

The French naturalist Pierre Belon rejected stories of fowl being generated so unnaturally; however, on the testimony of credulous friends, both Gesner and Ulisse Aldrovandi (ca. 1522–1605) accept tree geese in their encyclopedic works. Meanwhile, the tradition of tree-and-sea fowl became even more believable through an eyewitness account matching that of Giraldus. John Gerard eulogizes the "Goose tree, Barnacle tree, or the tree bearing Geese" in his *Herball* (1597):

> *There are found in the North parts of Scotland and the Islands adjacent, called Orchades, certaine trees whereon do grow certain shells of a white color tending to russet, wherein are contained little living creatures.*

Also, while walking on an English shore, he discovered on a rotted tree thousands of shells containing living things shaped like birds and covered with "soft downe." He avers that, "that which I have seene with mine eies, and handled with mine hands, I dare confidently avouch, and boldly put downe for verity."

Eyewitness reports and learned theories about the birth of geese continued into the mid-eighteenth century. Only after the study of the true nature of certain barnacles was the centuries-old fable concerning an actual bird—the barnacle goose *(Branta leucopsis)*—put to rest. One supposed source of the myth was linguistic confusion with shellfish. The French *canard,* meaning "hoax," "false report," was applied to the geese of the Orkney Islands and other northern sites.

A colored print of the generation of barnacle geese from John Gerard's late sixteenth-century Herball, or Generall Historie of Plants. *He ends his book with "this wonder of England. For the which Gods Name be ever honored and praised."*

The Polypus

Map D Ⓜ

Between the Orcades (Orkney Islands) and the Hebrides, a giant lobster grips a swimmer in its claws. The vignette to the *History* chapter (see left) on the Polypus depicts an equally gruesome attack on a sailor, pulling him off a ship down to its awaiting family. Most of Olaus's localized commentary is devoted to a natural-history description of a multi-legged creature—whatever it might be.

"On the Coasts of Norway there is a Polypus, or Creature with many feet, which hath a pipe on his back, whereby he puts to Sea, and he moves that sometimes to the right side, sometimes to the left. Moreover, with his Legs as it were by hollow places, dispersed here and there, and by his Toothed Nippers, he fastneth on every living Creature that comes near to him, that wants blood. Whatever he eats, he heaps up in the holes where he resides: Then he casts out the Skins, having eaten the flesh, and hunts after fishes that swim to them: Also he casts out the shels, and hard out-sides of Crabs that remain. He changeth his color by the color of the stone he sticks unto, especially when he is frighted at the sight of his Enemy, the Conger. He hath 4 great middle feet, and in all 8; a little body, which the great feet make amends for. He hath also some small feet that are shadowed, and can scarce be perceived. By these he sustains, moves, and defends himself, and takes hold of what is from him: and he lies on his back upon the stones, that he can scarce be gotten off, unless you put some stinking smell to him."

This many-footed creature is notable for the "pipe" on its back and its ability to change color. Olaus later cites authorities who maintain that the Polypus is a savage beast intent on killing swimmers and sailors.

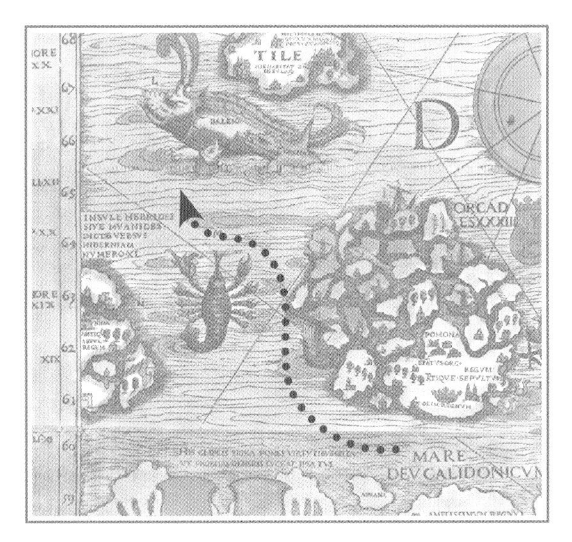

The course winds north to the west of the diocese of the thirty-three Orcades islands, which once comprised a kingdom. Between islands busy with shipping is a beached sea monster.

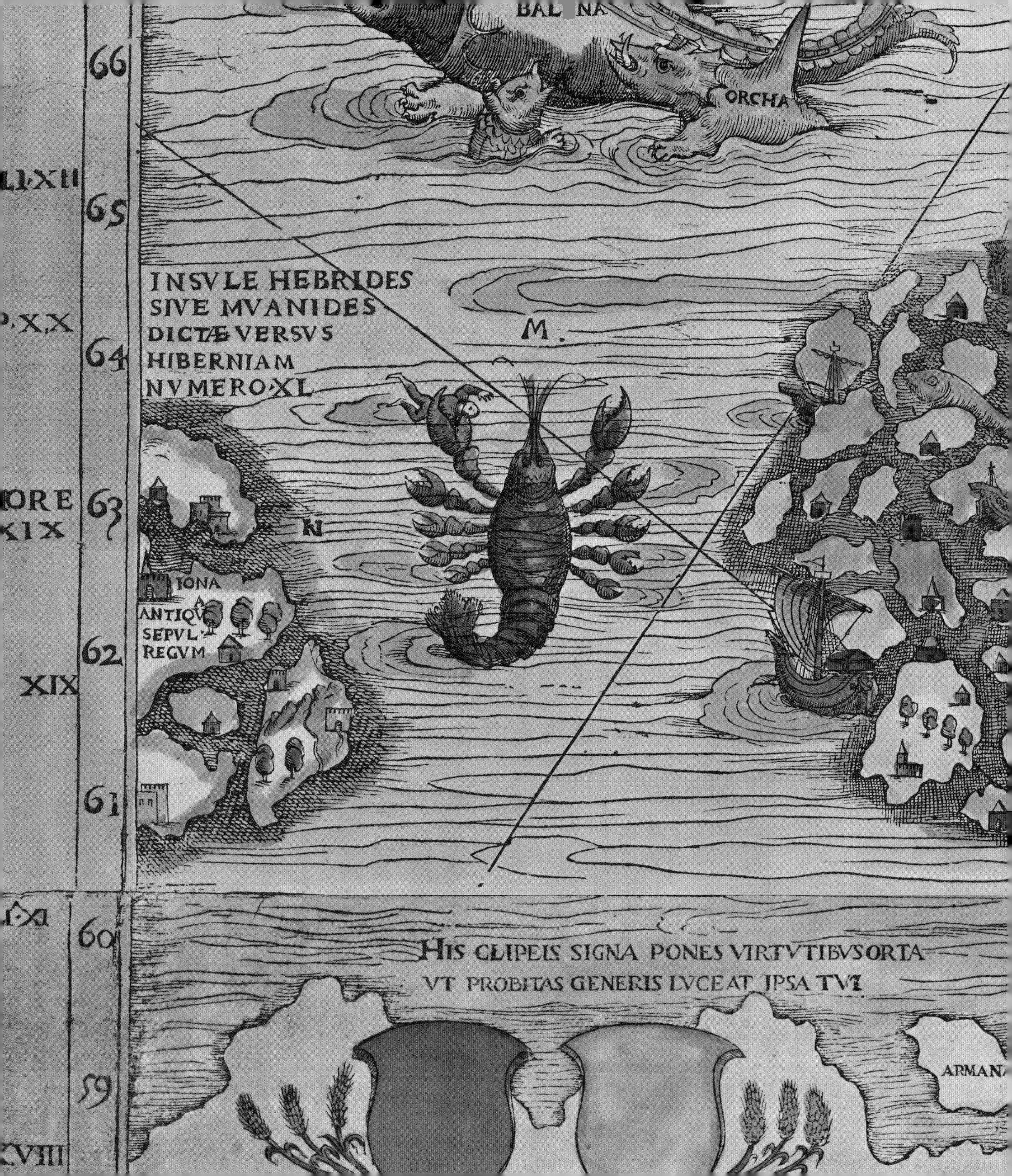
BAL NA
ORCHA
66
65
64
63
62
61
60
59
INSVLE HEBRIDES
SIVE MVANIDES
DICTÆ VERSVS
HIBERNIAM
NVMERO XL
M.
N
IONA
ANTIQV
SEPVL
REGVM
HIS CLIPEIS SIGNA PONES VIRTVTIBVS ORTA
VT PROBITAS GENERIS LVCEAT IPSA TVI
ARMAN
XIX

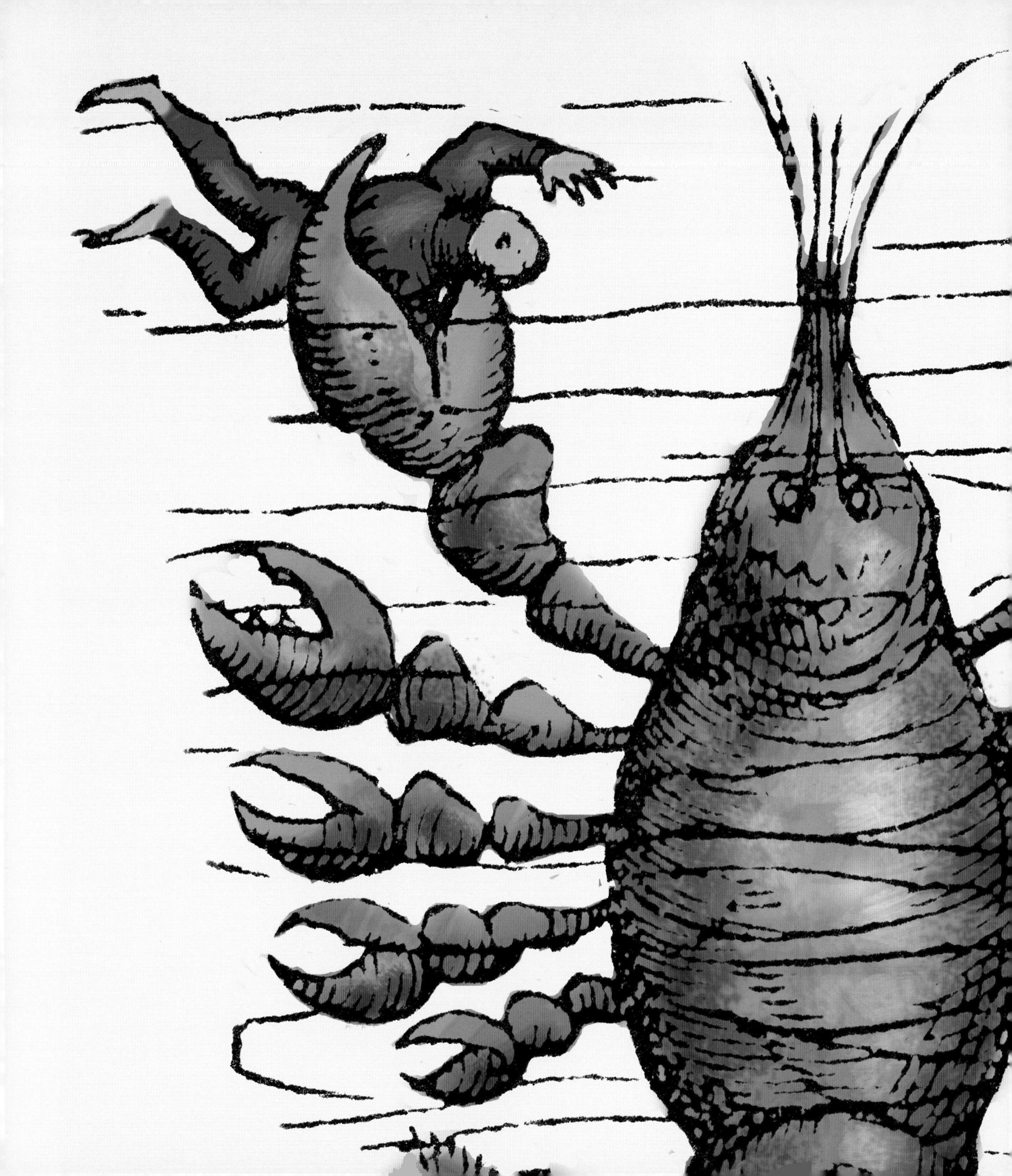

THE POLYPUS

Carta Marina

What animal does Olaus describe in his Polypus commentary? Is it the lobster depicted on the *Carta Marina* (D m) and in the *History* vignette? Not really. Nearing the end of years of labor on his mammoth book on Scandinavia, the learned scholiast does indeed discover a "polypus" in his standard authorities, but it is a different animal than the one he thought it was.

The lobster on the *Carta Marina* is true to its generic name. Polypus (meaning "many-footed") describes a great variety of animals, from the crab to the centipede. "On the Polypus" and "On the Octopus" are, in fact, the chapter titles in the older and the more recent English translations of the *History*. Both terms are literally accurate.

Because Olaus himself is thought to have either drawn or selected the figures on the *Carta Marina* and the vignettes in his *History*, it is clear that he had a lobster in mind when he placed the image on his map and in his book. It is also clear, however, that the sea beast he describes in his commentary is not a lobster.

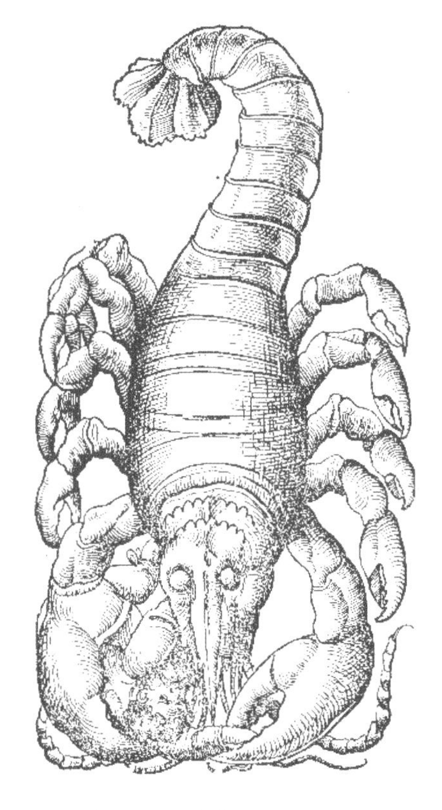

Conrad Gesner's chilling version of Olaus's Polypus, reproduced in Ulisse Aldrovandi's book of fishes.

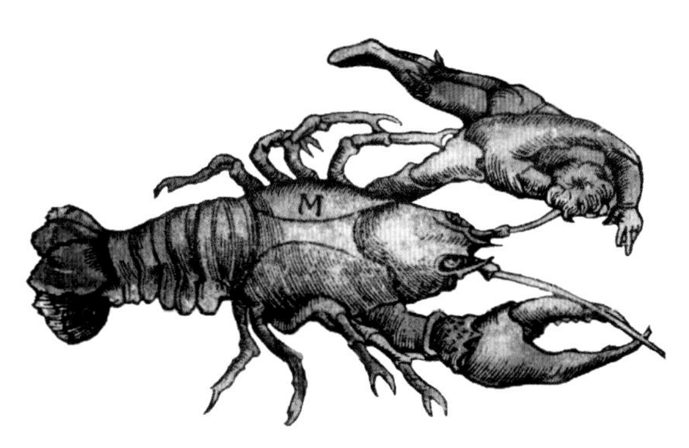

Sebastian Münster's "crab" (gambar) adaptation of the Carta Marina *lobster and swimmer.*

Ancestral Lore

Aristotle systematically distinguishes between mollusks and crustaceans, both of which include many-footed species. He states that the octopus has a "hollow tube" in front of its sac for discharging seawater that might have entered during feeding. Like Olaus's Polypus moving its "pipe" to the right and the left, this creature can shift its "tube" from one side to the other. In his *Natural History,* Pliny repeats this detail in his extensive description of the "polyp," an octopus.

Later in his *History* chapter, Olaus specifically cites Pliny and Albertus Magnus on the vicious nature of the Polypus. Pliny attributes the story to a Roman naturalist, Trebius Niger:

> *. . . no animal is more savage in causing the death of a man in the water: for it struggles with him by coiling round him and swallows him with its sucker-cups and drags him asunder by its multiple suction, when it attacks men that have been shipwrecked or are diving.*

More than a millennium later, Albertus Magnus refers to Pliny early in his entry on the "Polipus (Polyp, Multipes)": "Pliny adds that the polyp is sometimes powerful enough to drag a man off a boat." This line would serve as a caption to the vignette heading Olaus's chapter, notwithstanding that Pliny's attacking animal is an octopus or squid, with coils and "sucker-cups."

Elsewhere in his natural history, Albertus writes that "crab" is a generic term for crustaceans that include crayfish and lobsters. He declares that lobsters are so afraid of the octopus that sight of one makes them die of fright. Ironically, Olaus repeats the line in an earlier sea monster chapter.

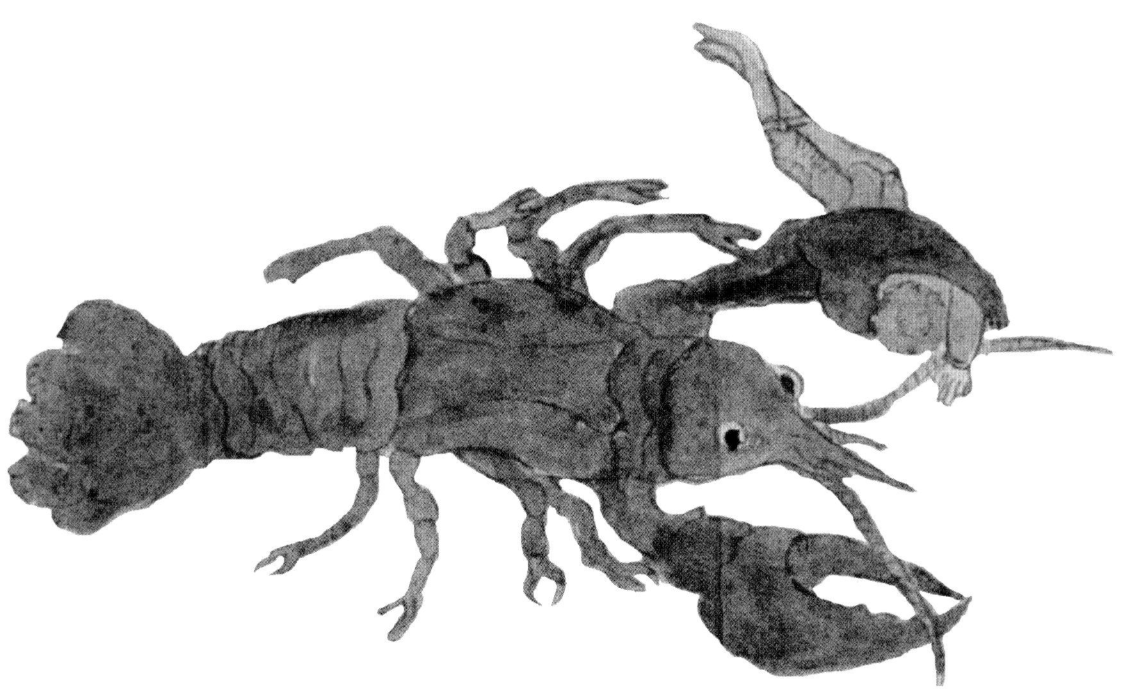

Adriaen Coenen's manuscript variation of the lobster on Sebastian Münster's Monstra Marina & Terrestria.

Map Legacy

Contemporaries who copied the *Carta Marina*'s lobster image—either from the map itself or secondhand from others—did not question the monster's crustacean identity. "Crab" and "lobster" being interchangeable at that time, they readily called it one or the other, unrelated to Olaus's *History* chapter. The image was compelling and the frightening action was believable because Pliny had recounted that spiny lobsters in the Indian Ocean were 6 to 8 feet long.

A woodcut in Walther Ryff's 1545 translation of Albertus Magnus's thirteenth-century natural history is reminiscent of Olaus's lobster. While both the monster and the victim in German pants differ markedly from the figures on the *Carta Marina*, the iconic scene is similar enough to suggest influence. Also, there are no man-eating lobsters in Albertus's book, which leads one to look to Olaus's map as the most likely source of inspiration for the image.

There is no doubt about Conrad Gesner's model for his 1560 lobster and swimmer woodcut. The caption reads: "Giant lobster catching a man; reported by Olaus Magnus in the seas between the Orkneys and the Hebrides." Ulisse Aldrovandi also uses this image in his book of fishes.

No giant lobsters appear on Abraham Ortelius's *Islandia*, but Sebastian Münster's engraver clearly adapted that chart's man-eating monster (M) from Olaus's map. As his key declares, such "Crabs of notable size" are so strong they "seize men with their claws and overwhelm them." Adriaen Coenen's 1585 watercolor copies Münster's image, in turn, and emphasizes the strength of the dangerous beast: "This large lobster can drag a swimmer down to the seabed with its claws." He adds somewhat rationalistically, that, "This is plausible, even if it applied to our ordinary lobsters."

A fanciful Victorian lobster primping before a mirror, from a John Tenniel engraving for Lewis Carroll's Alice's Adventures in Wonderland *(1865). As the lobster declares in a song: "You have baked me too brown. I must sugar my hair."*

And Since

Although naturalists of Olaus's time seemed unaffected by his juxtaposition of lobster images with octopus commentary, that is not true of at least one nineteenth-century author. Olaus's confusion resurfaces in John Ashton's cryptozoological classic, *Curious Creatures in Zoology* (1890). Among the many *Carta Marina* sea monsters Ashton discusses is Olaus's giant lobster/Polypus, which he includes in a section devoted to the Kraken. That great legendary beast, commonly equated with the giant squid, will not be encountered until very much later in the *Carta Marina* voyage.

BALENA & ORCA

Map D Ⓛ

The course bears toward the calm waters of a whale breeding ground off the island of Tile (Tyle). This *Carta Marina* island does not exist on other maps. It is commonly associated with the mythical Thule, an island the Greek explorer Pytheas (fourth century BC) reported in his voyage to the ends of the earth. Olaus describes the varieties of whales and the affection they have for their young:

"There are many kinds of Whales, some are hairy, and of four Acres in bigness: The Acre is 240 long, and 120 broad: some are smooth-skinned, and those are smaller, and are taken in the West and Northern Sea; some have their Jaws long and full of Teeth, namely 12 or 14 foot long, and the Teeth are 6 or 8, or 12 foot long. But their two Dog-Teeth, or Tushes, are longer than the rest, underneath like a Horn, like the Teeth of Bores, or Elephants. This kind of Whale hath a fit mouth to eat: and his eyes are so large, that 15 men may fit in the room of each of them, and sometimes 20 or more, as the Beast is in quantity. His Horns are 6 or 7 foot long, and he hath 250 upon each eye, as hard as Horn, that he can stir stiff or gentle, either before, or behind."

He adds that, *"They carry their young ones, when they are weak and feeble; and if they be small, they take them in at their mouths. This they do also when a Tempest is coming; and after the Tempest, they vomit them up."*

Then: *"A Whale is a very great fish about one hundred or three hundred foot long, and the body is a vast magnitude; yet the Orca, which is smaller in quantity, but more nimble to assault, and cruel to come on, is his deadly Enemy. An Orca is like a Hull turned inside outward; a Beast with fierce Teeth, with which, as with the Stern of a Ship, he rends the Whales Guts, and tears his Calves body, or he quickly runs and drives him up and down with his prickly back, that he makes him run to Fords* [fjords]*, and Shores."*

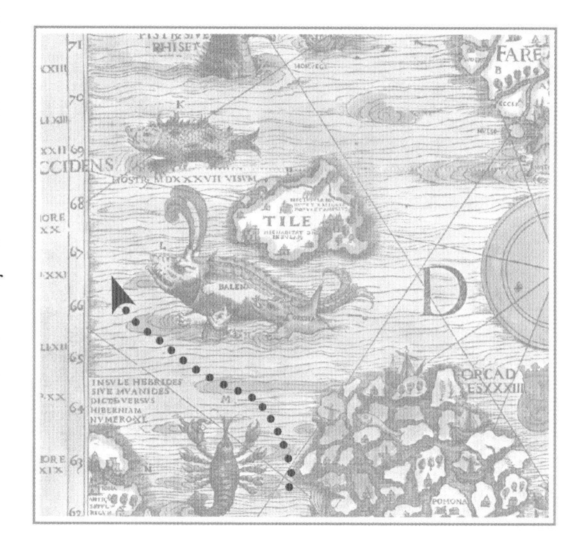

The island of Tile seems to be Olaus's invention. The Carta Marina *island boasts two castles and a legend that declares a population of thirty thousand.*

70
K
69
CIDENS
MOSTR MDXXXVII VISUM
68
HECINSVLA HA
BETXXXMILLIA
POPV: ETAMPLIVS
TILE
HICHABITAT DNS
INSVLAR
L
67
BALENA
66
ORCHA
65
INSVLE HEBRIDES
SIVE MVANIDES
DICTÆ VERSVS
HIBERNIAM
NVMERO·XL
M.
64
63
N

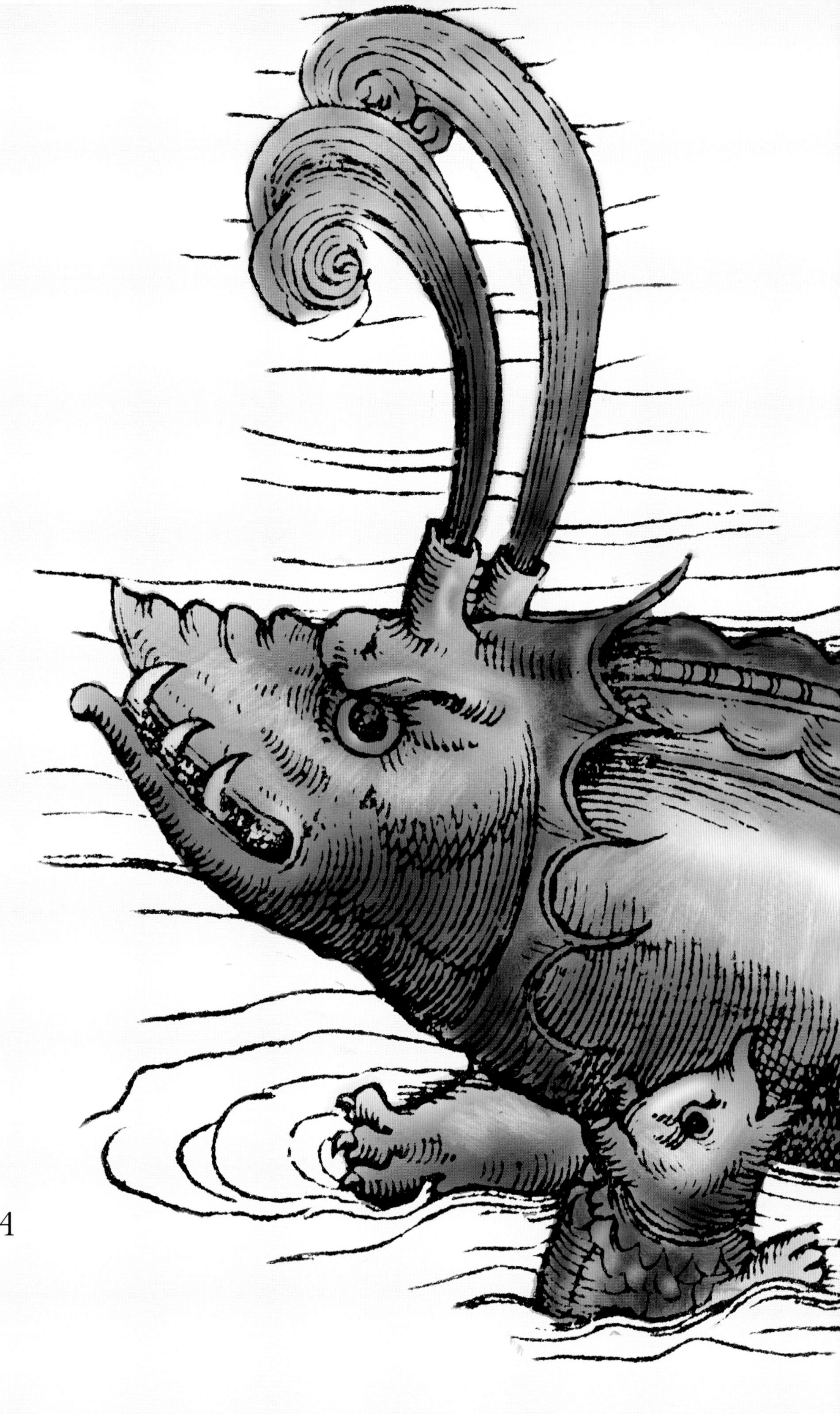

BALENA & ORCA

Carta Marina

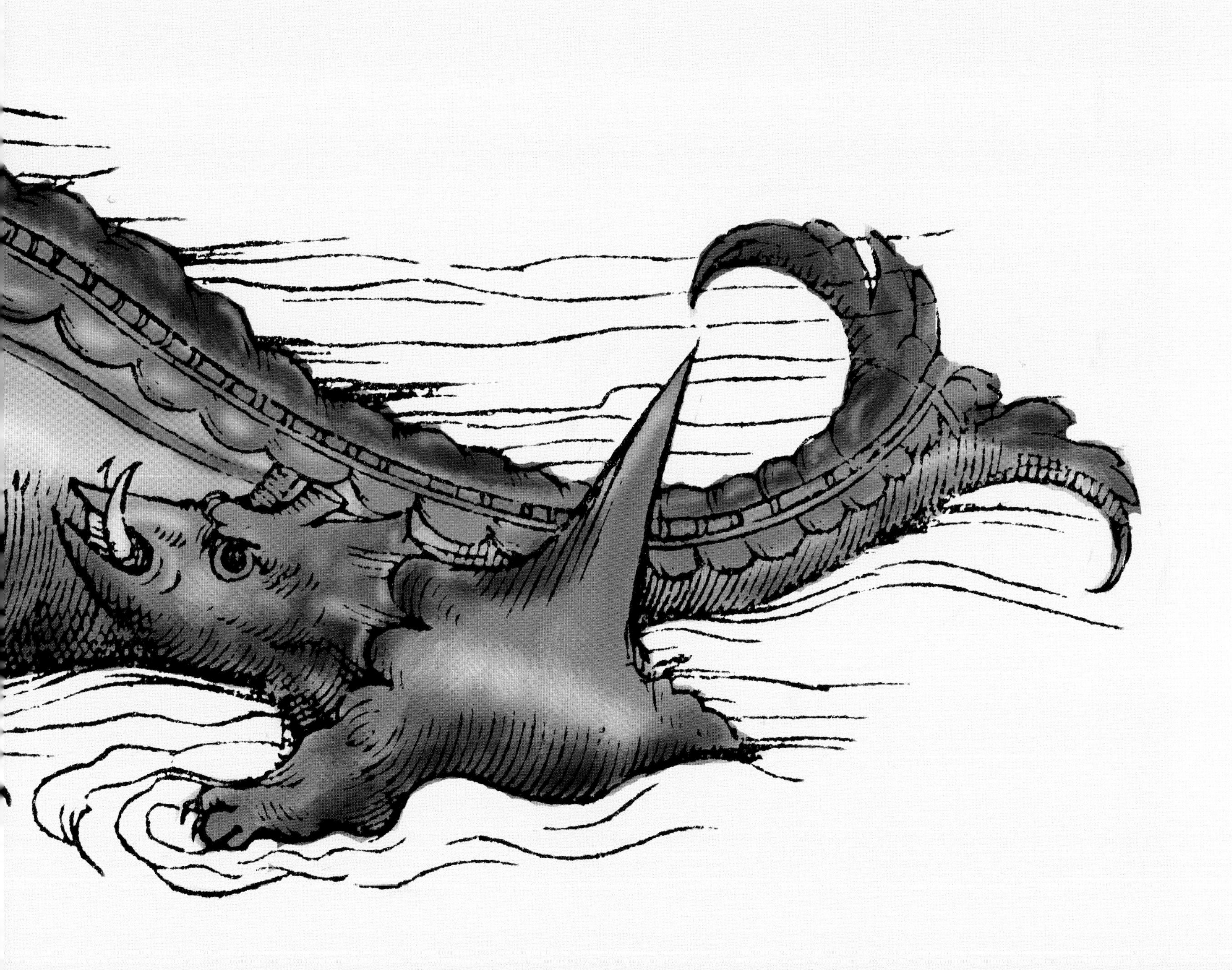

Like most of the marine figures on the *Carta Marina*, the first whale to be encountered on a course up the map is Olaus's attempt to depict an actual creature. The animal he calls "a great fish" is pictured nonetheless as a mammal with its calf. The image is in keeping with his description of whales' affection for their young and is one of the earliest illustrations of a nursing whale. The mother's spouting blowpipes, pointed snout, teeth, ruff around the neck, pawlike flippers, ridge down its back, and cetacean fluked tail place Olaus's Balena figure midway between great fish that medieval bestiaries presented as whales and natural history engravings based upon observation of beached whales. Variations of the image multiply on maps throughout the century and influence map decoration thereafter.

Ancestral Lore

Olaus uses the words "whale" and "monster" interchangeably in his map key and *History*. At one point, he refers to "monstrous fish" that have "unusual Names though they are reported a kind of Whales." Any large marine mammal was commonly called a "whale" in Olaus's time. Legend and folklore persisted along with the few natural history attempts to describe different kinds of whales.

Classical authors refer to Mediterranean whales, of which sightings were few, and not clearly represented in art. Aristotle maintains that "the dolphin, whale, and all the rest of the cetacea" differ from other "fishes" in having lungs and blowholes instead of gills and bear live young. The order of mammalian "Cetacean" comes from Latin *cetus,* from Greek *ketos,* "sea monster." The two standard suborders of Cetacea are baleen whales, with "horny" plates in their mouths, and the larger group of toothed whales, which include dolphins and porpoises. "Baleen" derives from Latin *balaena*, "whale" or other large marine mammal.

Olaus often consulted Aristotle's works, but in his commentary on the kinds of whales, he depends mostly on the natural history of Albertus Magnus. In his "Cetus" entry, Albertus writes that female whales are called "balaena," the term Olaus uses for his mother whale on the *Carta Marina*. James J. Scanlan, translator of Albertus's *De Animalibus*, notes that the animal Albertus describes as having tusks like boars or elephants is actually a walrus. Olaus repeats the error as well as

Sebastian Münster's orca by itself, without the Carta Marina*'s Balena and calf. This beast, with spouts, tusk, and paws, is clearly an Olaus Magnus whale. With its sharp, pointed snout and fin, it appears even more vicious than its* Carta Marina *prototype.*

A colored print of Conrad Gesner's copy of the Carta Marina*'s Balena, calf, and Orca. Like perhaps all Gesner woodcuts of Olaus's figures, the image is from the map, not a* History *vignette, and is reversed in the printing process.*

additional inaccuracies that Scanlan points out: that the whale sockets that Albertus affirms can accommodate up to twenty men are not eyes, but mouths, and that "horny appendages" above the eyes are baleen plates. Scanlan concludes that Albertus's description most closely matches that of the Greenland right whale (bowhead whale), whose baleen plates of 10–15 feet in length number about 300 in each side of its mouth. As this voyage shows, the transmission of inaccurate or misinterpreted authority is hardly uncommon among men of learning in such a transitional time.

Albertus's entry on the "Orcha" briefly echoes Pliny, the first to describe the killer whale. Olaus draws from Pliny's longer sensational account of the grampus as an "enormous mass of flesh with savage teeth." The vicious beasts charge females and calves "like warships ramming" and bite and mangle them as the sluggish animals vainly try to escape. Pliny goes on to recount a battle between such a beast and the Roman emperor Claudius (reigned AD 41–54) at the harbor of Ostia. One imperial ship was sunk when soldiers trapped the monster behind nets and speared it to death.

Map Legacy

Sebastian Münster echoes Olaus in affirming that Norwegians call the orca *springval*—"because of its sprightliness, and it has a high, wide hump on its back." The orca depicted on Münster's *Monstra Marina* (L) does have the "hump" of a dorsal fin, but is pictured without the whale and calf.

The Balena is also absent from Abraham Ortelius's *Islandia*. An Orca counterpart would seem to be his "Staukul" (M), which is called *Springval* in Dutch because of its leaping. It is depicted and described differently from either Olaus's or Münster's beast.

Conrad Gesner reproduces Olaus's poignant attack scene in one of his numerous whale woodcuts, and Adriaen Coenen accompanies his watercolor version of the figures with, "Olaus Magnus describes the struggle between the whale and the orca in his 21st book."

Adriaen Coenen's own version of Olaus's spouting mother whale, with fishlike fins and clawed paws. Coenen's pictures of Olaus's monsters are usually derived from Gesner woodcuts, Sebastian Münster's chart, or Olaus's History *vignettes, not the* Carta Marina.

And Since

The name and the teeth of Olaus's Balena evoke both major orders of whales: baleen whales, *Mysticetsis* (from Greek meaning "moustache" and Greek/Latin "whale" or "sea monster"), and *Odontocetes* (from Greek/Latin), toothed whales. Its real-life prototype could be the endangered northern right whale *(Eubalaena glacialis)*, whose natural predator is thought to be the orca. Olaus's and Münster's Orca, with its distinguishing dorsal fin, approximates the toothed killer whale, which is the largest of the cetacean family of dolphins.

The Sea Swine

Map D Ⓚ

A prevailing belief from classical times through to Olaus's own was that all living things on the land had their counterparts in the sea. One of those is the Sea Swine, which the *Carta Marina* key identifies as, "A sea monster similar to a pig." Such a creature was one of two monsters discovered on the shores of what is now known as the North Sea. Olaus describes the other one first:

"[A] monstrous Fish found on the shores of England, with a clear description of his whole body, and every member thereof, . . . was seen there in the year 1532, and the Inhabitants made a prey of it. Now I shall revive the memory of that monstrous Hog that was found afterwards, Anno 1537, in the same German Ocean, and it was a Monster in every part of it. For it had a Hogs head, and a quarter of a Circle, like the Moon, in the hinder part of its head, four feet like a Dragons, two eyes on both sides of his Loyns, and a third in his belly inkling toward his Navel; behind he had a Forked-Tail, like to other Fish commonly."

Olaus continues, explaining that an interpretation of that "monstrous Hog" was printed in Rome soon after discovery of the beast. The author likened the creature to heretics, who lived like swine. The misplaced moon signified their distortion of truth, the eyes on its scaly body their temptations to others, and its dragonlike feet the evil that they spread throughout the world. Altogether, the monster served as an example to unclean people to change their lives, to cease their monstrous ways, and embrace goodness. Olaus scornfully concludes that the sea pig is a savage beast that, like other predators, preys on the weak and the feeble.

The northern course along the far western edge of the map passes the snarling beast with the Roman numeral legend of the year it was first sighted.

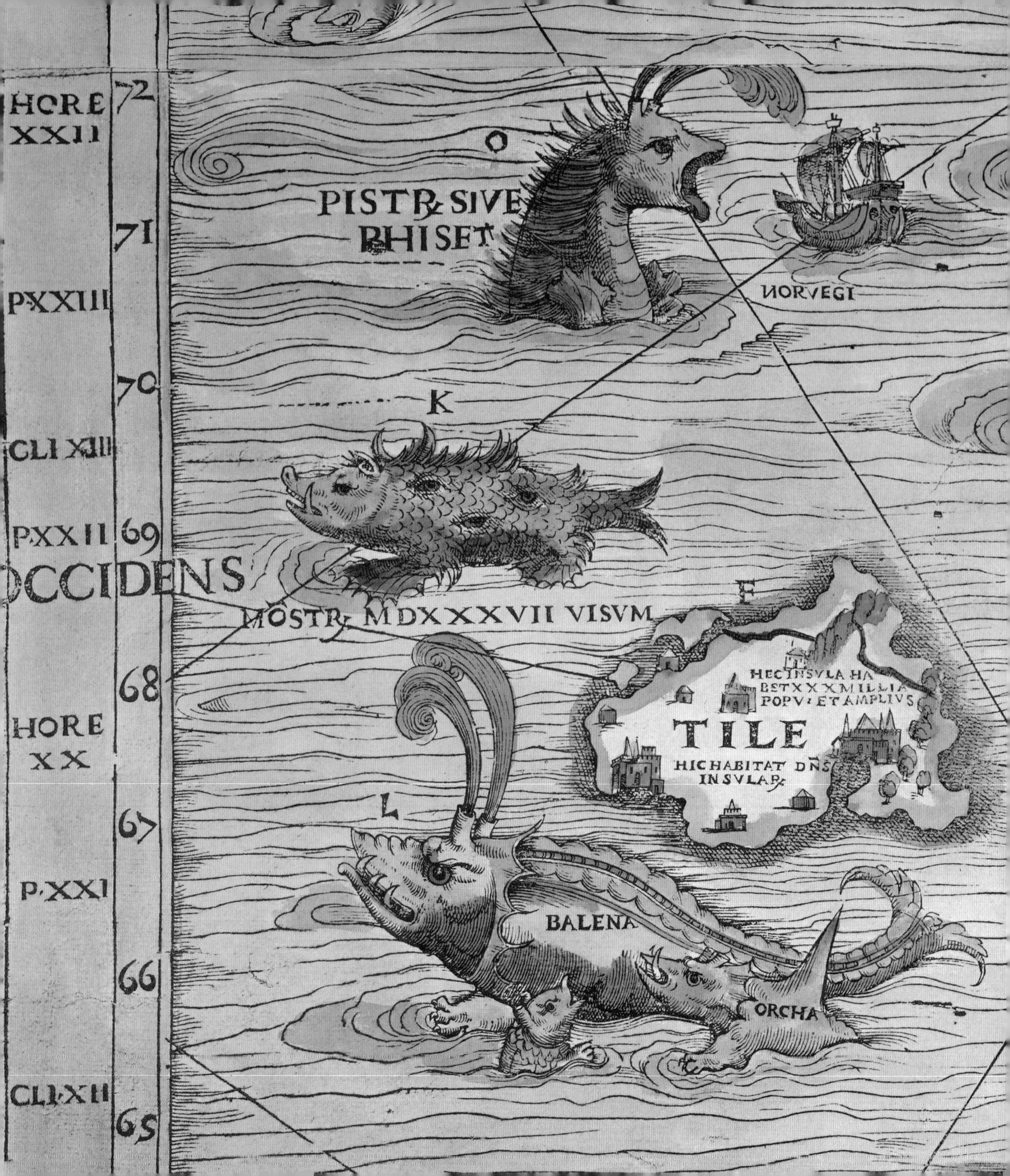

HORE XXII
72
O
PISTRX SIVE PHISET
71
NORVEGI
P·XXIII
70
K
CLI XIII
P·XXII
69
OCCIDENS
MOSTR MDXXXVII VISUM
F
68
HEC INSULA HA BET XXX MILLIA POPU ET AMPLIUS
TILE
HIC HABITAT DNS INSULAR
HORE XX
L
67
P·XXI
BALENA
66
ORCHA
CLI·XII
65

SEA SWINE

Carta Marina

Olaus's description of the Sea Swine is not typical of his commentaries on the sea monsters of *Carta Marina*. Instead of drawing heavily upon classical and medieval learning, he derives his sea beast from an account that was printed only two years before he completed his map.

The source of the obscure allegorical description of the "monstrous Hog" was a pamphlet, *Monstrum in Oceano Germanico,* with "a theological interpretation." It was published in Rome on October 15, 1537. Olaus is sometimes named as the author of the anonymous pamphlet, but there is no solid evidence to substantiate that claim. As John Granlund relates in his article on the *Carta Marina*, a rare copy of the previously unknown document was discovered in 1945.

The extension of Olaus's commentary on the Sea Swine is a Counter Reformation attack on Protestants. The allegory would have special meaning for him, a Catholic ecclesiastic in exile from his native Sweden. While the commentary is more openly didactic than his others on sea monsters, the bitter message is in keeping with his religious and nationalistic purposes in creating the map and writing the *History*.

Ancestral Lore

Olaus's Sea Swine source purported to be based on an actual animal captured in the North Sea. The fantastic allegory gives little hint as to the identity of a real creature except

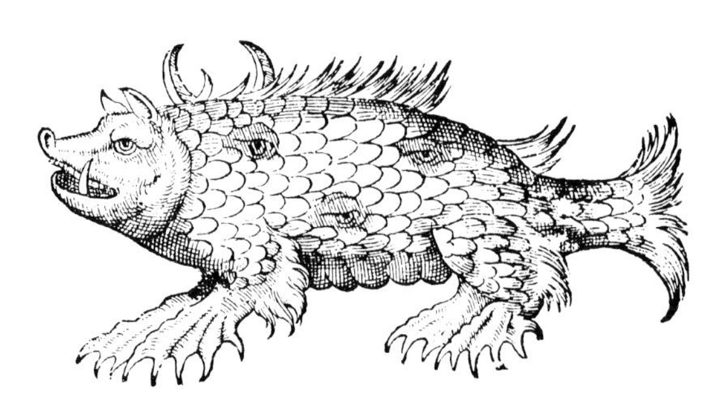

Ambroise Paré's "monstrous Sea-Swine," from Conrad Gesner's woodcut. Paré, like Gesner, attributes the beast to Olaus, but the French surgeon mistakenly reports it was discovered in 1538.

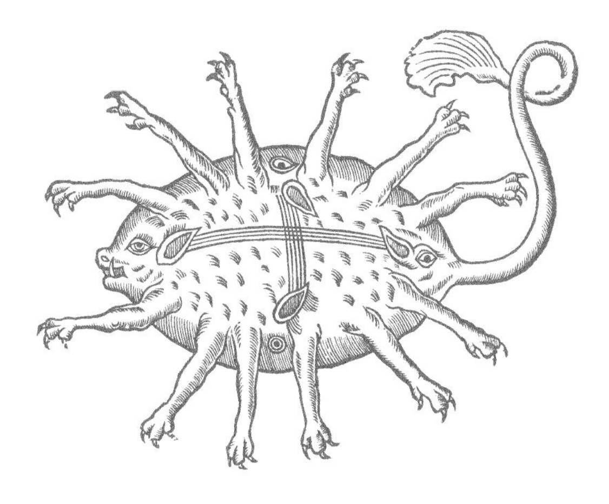

Another piglike creature from Conrad Gesner's natural histories. This fantastic twelve-legged figure with pig's ears and eyes has a body marked like a hot cross bun and a looping tailed nose like a monster in a Hieronymus Bosch painting. It was reported sighted between Nice and Antibes in 1562.

to compare it to swine. There is a long tradition of writings on the "sea-pig" *(Porcis marinus)*, marine animals with perceived characteristics of swine.

Pliny records that the "pig-fish" grunts like a swine when it is caught and that the Lacedaemonians called it *orthagoriscus,* which is an ocean sunfish. He also writes that poisonous spines on the back of a "sea-pig" cause great pain until a wounded person drinks bile from the fish's body. Albertus Magnus cites Pliny's observations in his "Porcus Marinus" entry. He adds that the head and other parts resemble those of a pig and that the beasts root in mud on the seafloor.

The *Porcus Marinus* text of the twelfth-century *Aberdeen Bestiary* continues the tradition, stating that, "Sea-pigs are commonly called swine, because when they seek food, they dig under the water like swine digging into the ground." Transmission of sea-pig lore continues in the encyclopedic *Hortus Sanitatis* (1491) and its Dutch translation, *The Palace of Animals* (1520).

Olaus would have been aware of such classical and medieval authorities; however, the pamphlet's allegory offered him the opportunity to express his negative views of Sweden's conversion to Lutheranism.

Map Legacy

Olaus's Sea Swine figure, adapted from the title page of *Monstrum in Oceano Germanico,* is a fantastic composite in the tradition of phantasmagorical medieval paintings. Even though it is doubtful that the *Carta Marina* monster is Olaus's own creation, it is one of his most frequently copied icons. On maps and in natural history illustrations, the sea swine is presented as the beast that Olaus pictures on the *Carta Marina* and describes in his *History*. There is no mention of the religious diatribe.

Sebastian Münster describes the creature in his chart's key entry (K) as, "Sea monster, somewhat resembling a pig, was seen in AD 1537." Abraham Ortelius writes in his key that, "The Hyena or sea hog is a monstrous kind of fish about which you may read in the 21st book of Olaus Magnus" (D). The obsolete meaning of "hyena" is "a ravenous fish," which is derived from a Greek word for "hog" or "pig."

Conrad Gesner's woodcut figure of Olaus's "monstrous hog" is identified as a "boarlike whale," which was sighted north of the Orkneys in 1537, according to Olaus Magnus. In his *Of Monsters and Prodigies*, Ambroise Paré reproduces a virtually identical likeness of Gesner's "boarlike whale," and repeats Gesner's detailed description of the animal's extravagant size. In the process, the year the monster was discovered inadvertently changes from 1537:

> *Olaus Magnus writes that this monster was taken at Thyle an Iland of the North, Anno Dom. 1538. It was of a bignesse almost incredible, as that which was seventy two foot long, and fourteene high, and seven foot betweene the eyes: now the liver was so large that therewith they filled five hogsheads.*

Sebastian Münster's piglike monster (K) adapted from the Carta Marina *figure. The eyelike designs in the beast's side outnumber the three eyes on the bodies of Olaus's, Gesner's, Paré's, and Coenen's monsters.*

Adriaen Coenen's delightful watercolor *zeevardkens* (sea swine) have all the charm of illustrations in a children's book. Paired with the sea pig from *The Palace of Animals*, his friendly version of Olaus's Sea Swine is modeled on the *History*'s "monstrous hog" vignette. His calligraphic labels attribute both the illustration and his descriptive text to Olaus. In a typically personal aside, he ends his manuscript text with, "I wrote this on Martinmas 1584."

Adriaen Coenen's whimsical rendering of Olaus's Sea Swine. It is clear by the number of eyes in the beast's body that Adriaen did not derive his figure from Münster's Monstra Marina & Terrestria, *as he often did.*

And Since

What actual marine mammal (or mammals) could have been the prototype for the *Monstrum in Oceano Germanico*'s fantastic, allegorical sea swine is certainly problematic. Traditional lore that sea swine root in the sea mud and Gesner's calling it a "boarlike whale" suggest similarities to the walrus. On the other hand, "sea swine," "swine-fish," "hog-fish," and "sea hog" all become terms for the porpoise early in the next century. The fantastic Sea Swine could, of course, be a blend of both—or many others, too.

THE SEA UNICORN

Map A *(Not Keyed)*

A straight pointed horn rises from the far northern waters, followed by the head of a sea monster with an elongated snout. While the horn of such a beast was sought for the high prices it brought in northern trading, Olaus draws on an earlier tradition to describe the dangers the animal posed to mariners. Disaster could be avoided by keeping a safe distance from the creature.

"The Unicorn is a Sea-Beast, having in his Fore-head a very great Horn, wherewith he can penetrate and destroy the ships in his way, and drown multitudes of men. But divine goodnesse hath provided for the safety of Marriners herein; for though he be a very fierce Creature, yet is he very slow, that such as fear his coming may fly from him."

The Sea Unicorn evokes yet another of the *Carta Marina*'s most pervasive and controversial medieval beliefs regarding the nature of certain animals. The Duck Tree viewed earlier touched on the possible miraculous birth of ducks and geese from trees or ocean barnacles. The sea beast with a single horn is the marine counterpart of the fabled terrestrial unicorn, whose long spiraled horn was prized for its magical medicinal properties.

The Sea Unicorn shares the vignette (above left) with a flying fish and other marine creatures. Olaus writes elsewhere in his *History* that Pliny called a fish with two wings a "sea swallow," which flies only briefly before diving back into the sea. The winged fish in the upper left corner of the vignette might be a flying gurnard. It would seem to be what Olaus mistakenly calls a "cuttle-fish," derived from Pliny's description of flying polyps.

The northern course holds steady past the Sea Unicorn, then suddenly bears east to avoid a sea battle raging between two ships competing for trade in Iceland. A cannonade from a Hamburg vessel cracks the mainmast of a Scottish ship, sending a sailor overboard. The Olaus voyage changes course to the southeast, beneath a flying fish with spiny wings.

HOR:XX
76
M
P XXV
75
HAMBVRGEN
SCOTI
CLI XIIII
74
P XXIIII
73
HORE
XXII
72
O
PISTRX SIVE
PHISETR
71
NORVEGI
P.XXIII
70
K
CLI XIII
P.XXII
69
OCCIDENS
MOSTRX MDXXXVII VISVM

Like other marine figures on the *Carta Marina,* the Sea Unicorn was meant to represent an actual animal. And it does—even though the imagination created it from hearsay instead of from observation. Because the picture is so small, has no map legend, and is not included in the map's key, it's clear that Olaus did not regard it as very important. Nonetheless, his sea beast with the single horn protruding from its forehead is perhaps the first printed rendering of a narwhal *(Monodon monoceros).* Olaus's commentary, too, is an early description of that cetacean of the Arctic seas. However, it is not the first. It echoes the "Monoceros" entry in Albertus Magnus's natural history three centuries earlier. Neither Albertus nor Olaus could have foreseen that the single-horned marine creature would sustain centuries of belief in the mythical unicorn and its storied medicinal horn.

Ancestral Lore

The oldest literary ancestor of the fabled unicorn is the wild ass of India. As Ctesias the Cynidian describes the beast in his *Indica* (fifth century BC), it is remarkable for having on its forehead a single black, white, and crimson-tipped horn about a forearm in length. He declares that when the horn is powdered or fashioned into drinking cups, it is an antidote to poison and prevents disease. Ctesias's creature is most often identified with the rhinoceros.

"The Unicorn in Captivity," the seventh and final panel of the famed Unicorn Tapestries *(Netherlands, ca. 1495–1505). The tapestries illustrate the spiraled narwhal tooth as the horn of the mythical unicorn.*

The "Nahval" from Abraham Ortelius's Islandia. *A narwhal counterpart of Olaus's Sea Unicorn, portrayed with characteristics of cartographical whales.*

The similar wild and ferocious Hebrew *re'em* of Judeo-Christian scripture transformed into the Greek *monoceros* and the Latin *unicornis.* Through the *Physiologus* and writings of the Church Fathers, the unicorn became the personification of the defiant and sacrificed Christ, the divine Son of God made human in the lap of the Virgin Mary. Olaus's source for the description of the narwhal was the ecclesiastic naturalist Albertus Magnus, who wrote about both the religious unicorn and the "Monoceros" marine animal.

The unicorn's spiraled horn had entered Christian art a century before Albertus, and it becomes the standard distinguishing feature of the graceful animal as it so famously appears in the more secular *Hunt of the Unicorn* and *Lady and the Unicorn* tapestries (Netherlands, ca. 1495–1505). Both are part of the virgin-capture tradition initiated in the *Physiologus* with the Virgin Mary. The only single "horn" in nature that is similar to those fabulous horns is that of the narwhal—which medieval artists could have seen.

A history of Iceland records that such horns were discovered by villagers among cargo washed up on shore after an 1126 shipwreck. The apparent value of the narwhal teeth, each inscribed with a sailor's name, suggested an ongoing European trade. Worth several times their weight in gold, narwhal teeth were sought for royal and ecclesiastical treasuries.

Natural history views of a male narwhal, from William Scoresby's An Account of the Arctic Regions with a History and Description of the Northern Whale Fishery *(1820).*

Map Legacy

Sebastian Münster does not include a Sea Unicorn on his *Monstra Marina & Terrestria,* but Abraham Ortelius does in his 1584 *Islandia* (A). Unlike Olaus's figure, the spouting beast with conventional whalelike ruff has a sharp, elongated snout representing a "horn." Ortelius's key entry, too, differs surprisingly from Olaus's (and Albertus's). He identifies the animal by its now-standard name: "NAHVAL" (Dutch for "narwhal") and states that its 7-cubit-long "tooth" is sometimes sold as "the unicorn's horn" for its medicinal powers. (Gerard Mercator, who often collaborated with Ortelius, repeats this passage in his 1621 *Atlas Minor.*)

The influence of Olaus's Sea Unicorn on Conrad Gesner was more direct than it was on the *Islandia* narwhal. Gesner's 1558 woodcut of the creature is a virtually identical copy of Olaus's figure. It is one of two kinds of "unicorns" in Gesner's natural histories. The other is the iconic equine unicorn, a creature whose existence and the power of its curative horn he accepted.

Shortly after Gesner's unicorn is famously reproduced in Edward Topsell's *History of Four-Footed Beasts* (1607), Gesner's copy of Olaus's Sea Unicorn surfaces unexpectedly in a painting commissioned by the eccentric Rudolf II (1552–1612), Holy Roman Emperor. The artist creates a rare vision of two kinds of sea unicorns—a fanciful unicorn hippocampus and Olaus's narwhal—together in the sea. That Rudolf II believed in unicorns is evident from his many royal treasures containing "unicorn horn." His favorite of all was a 6-foot narwhal tooth. When in the depths of melancholy, he would take it and an agate bowl that he believed was the Holy Grail and draw a circle around himself with a Spanish sword for protection from his enemies.

Conrad Gesner's "monocerote," the often reproduced version of the fabulous unicorn of medieval bestiaries and tapestries. Gesner's woodcuts of Albrecht Dürer's rhinoceros and Olaus's Sea Unicorn also featured single-horned animals.

And Since

One of the most famous of all unicorn horns is the "Horn of Windsor," which Martin Frobisher presented to Queen Elizabeth I upon return from his second voyage in search of the Northwest Passage in 1577. His men had discovered a "dead fish," an entire narwhal, whose spiraled hollow tooth was 2 inches shy of 2 yards. They declared the tusk a true unicorn horn. The royal treasure was later valued—accurately or not—at £100,000.

Danish physician and collector Ole Worm revealed to Copenhagen apothecaries in 1638 that what they considered to be unicorn horn was the tusk or tooth of a narwhal. That revelation barely dampened belief in the curative powers of the unicorn, which persist into the eighteenth century.

THE PRISTER

Map D Ⓞ

A cloud of mist swirls over the western waters of the *Carta Marina*. The source of the waterspout is not the sea itself, but a monster rising up out of the waves. It is a towering horselike figure with a ribbed, dragonlike chest, threatening to sink a ship with its torrential spouting. Olaus describes such beasts that terrorize mariners of the northern seas:

"The Whirlpool, or Prister, is of the kind of Whales, two hundred Cubits long, and is very cruel. For to the danger of Sea-men, he will sometimes raise himself beyond the Sail-yards, and casts such floods of Waters above his head, which he had sucked in, that with a Cloud of them, he will often sink the strongest ships, or expose the Marriners to extream danger."

What kind of whale is Olaus's "Prister?" Only its spouting is similar to that of the mother Balena encountered earlier. As with other monsters on the *Carta Marina*, this sea beast is known by many names. The map legend accompanying the figure is equivalent to "pister [also pristes]" or "physeter," sea monster names that Olaus accepted as interchangeable on his map and in his Latin *Historia*.

The "Whirlpool" and "Prister" of Olaus's commentary come from the 1658 abridged *Compendious History of the Goths, Swedes, and Vandals*. The 1998 English translation of Olaus's voluminous work, *A Description of the Northern Peoples*, renders his text as "spouter" and "leviathan." The Renaissance translator's "Prister" is an obscure Anglicized construction of the Latin *Historia's* "pristes." The remainder of the 1658 "Prister" chapter will supply Olaus's commentaries for forthcoming *Carta Marina* whale images.

The voyage course skirts the endangered Carta Marina *ship that the legend identifies as Norwegian and bears southeast toward the Faroe Islands.*

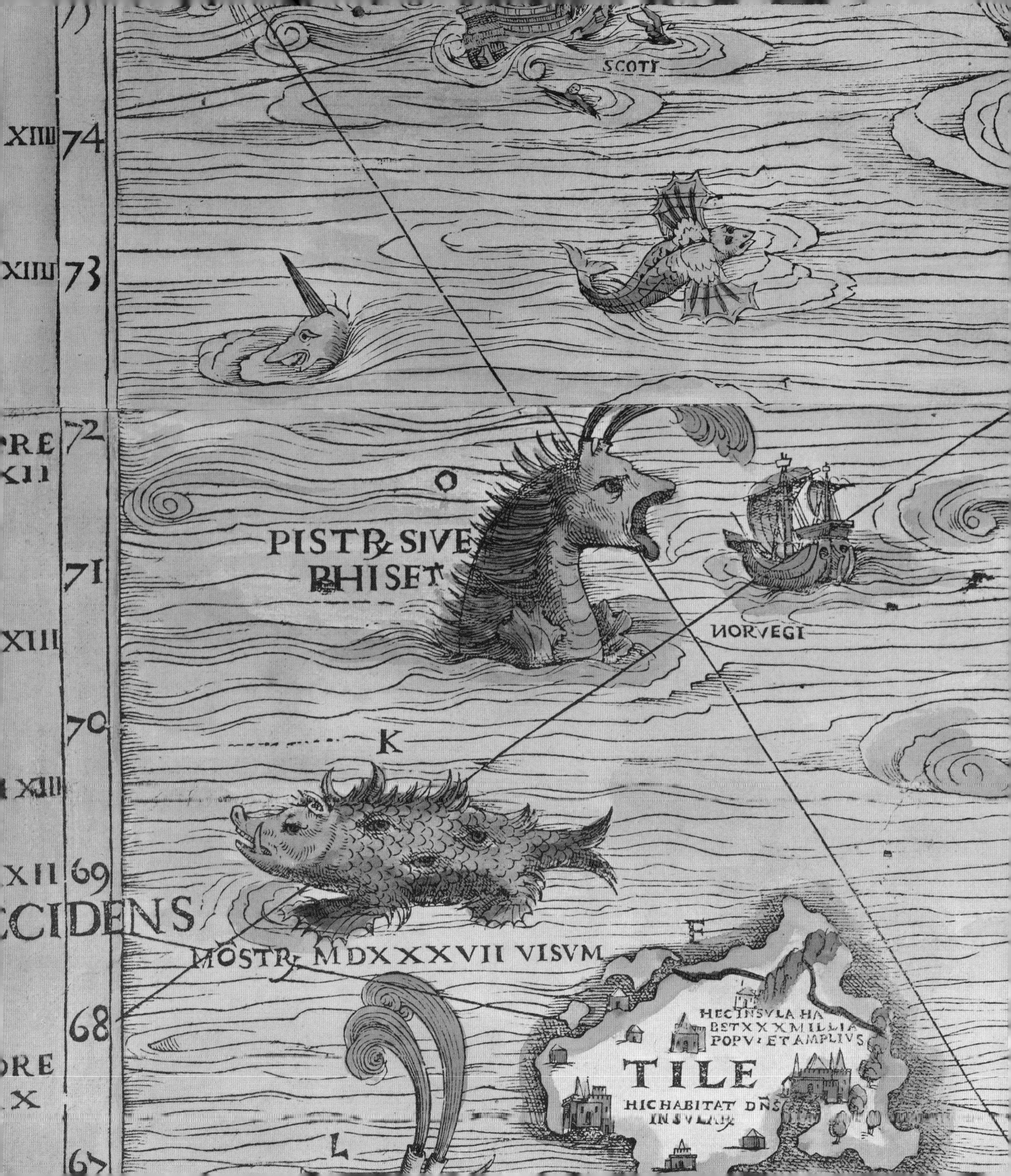
SCOTI
XIIII 74
XIIII 73
72
RE
XII
O
PISTR SIVE
RHISET
71
NORVEGI
XIII
70
K
XIII
XII 69
CCIDENS
E
MOSTR MDXXXVII VISVM
68
HEC INSVLA HA
BET XXX MILLIA
POPVL ET AMPLIVS
TILE
HIC HABITAT DNS
INSVLAE
ORE
X
L

THE PRISTER

Carta Marina

Sebastian Münster's *spouting "physeter," with horselike mane. Conrad Gesner and Gerard Mercator also adapt the Prister/Physeter image from Olaus's* Carta Marina.

The *Carta Marina* Prister, which dwarfs a fragile ship and engulfs it with its torrent of seawater, is a powerful, iconic figure born of the imagination. The monster's great horselike head, lolling tongue, bristles down its back, and ribbed abdomen are those of some variant of what could be called a sea dragon. Fantastic though the picture is, it is meant to represent an actual marine creature, "of the kind of Whales." Apart from the spouting beast's traditional lore, the figure has a pictorial legacy all its own.

Ancestral Lore

The literary inspiration for Olaus's spouter figure can be traced back through one of his major medieval sources, Albertus Magnus, to the popular classical authority Pliny the Elder. The Roman encyclopedist and naturalist wrote that:

> *The largest animals in the Indian Ocean are the shark* [pristis] *and the whale* [balaena]*; the largest in the Bay of Biscay is the sperm whale* [physeter], *which rears up like a vast pillar higher that a ship's rigging and belches out a sort of deluge.*

Translator H. Rackham's "shark" comes from the Latin word that also means "whale," "saw-fish," "any sea monster," and was a common Roman name for the constellation Cetus, or the Whale. The generic *balaena* now denotes baleen whales. *Physeter,* from Greek for a large whale that is a "blower" or a "spouter," refers specifically to the sperm whale, whose scientific name is *Physeter macrocephalus*. The largest of the toothed whales, it is also known as the "cachalot," (from archaic French, "tooth"). Contrary to tradition, the exhaled blow of cetaceans is primarily condensed vapor, not water. The sperm whale and other toothed whales have only one blowhole; baleen whales have two.

An emblem of a spouting whale with tusks, ruff, and spouts in the tradition of Carta Marina *whales other than the horselike Prister image, from the 1604* Symbolorum et emblematum *of fish and reptiles by German scholar Joachim Camerarius the Younger (1534–1598).*

Pliny and Albertus are among the few naturalists who attempt to categorize the species of marine life prior to the zoologists of the sixteenth century. However, while Albertus was compiling his natural history, the anonymous author of an Old Norwegian manuscript names and describes the many kinds of Icelandic whales. These range from the grampus and the "beaked," "hog," and "raven" whale to the rorqual. There is no indication that Olaus—or his contemporaries—were aware of *The King's Mirror* (ca. 1250), a dialogue in which a father instructs his son on the wonders of the world. Derived from observation and local lore, the description of the sperm whale differs markedly from scholarly tradition. Such whales are "neither fierce nor savage, but rather of a gentle nature, and so far as possible they avoid the fishermen." More than seventy teeth in a single whale "are barely large enough to be carved into fair-sized knife handles or chess men."

Map Legacy

Shortly after the printing of the *Carta Marina*, an imitation of Olaus's Prister image surfaces in a surprising way on Gerard Mercator's 1541 terrestrial globe. The beast rises out of rhumb lines between a cartouche and a pair of dividers, accompanied by a legend that identifies it as "Hippopotamus." The literal "river horse" actually describes the unusual equine image more accurately than *Carta Marina*'s Prister legend. Ten years later, Mercator presented his vision of a whale—totally different from a hippopotamus and Olaus's Prister—on the celestial globe of his matching pair. Among the constellations is Cetus, the Whale, whose legend includes "Pistrix" and "Balena."

The difference between Olaus's horselike figure and cartographical whales is dramatically illustrated on Sebastian Münster's *Monstra Marina & Terrestria*. The fantastic beast that Münster describes as a "terrifying type of huge monsters which are called physeteres" (B) blows a deluge of water onto a crowded ship, blocking its progress, while an indisputable whale (A) menaces the vessel in its wake. Münster identifies his chart's two whale figures as "Great whales as large as mountains."

The Prister of the *Carta Marina* does not appear on Abraham Ortelius's *Islandia*, but several scholars have suggested that the image inspired the "physetere" in the 1552 volume of François Rabelais's raucous mock-epic, *Gargantua and Pantagruel*.

A sperm whale on a stamp of the Republic of Guinea-Bissau, in West Africa (1984).

"The monstrous physetere was slain by Pantagruel."

Gustave Doré's nineteenth-century illustration of the physeter chapter in Rabelais's Gargantua and Pantagruel. *Doré portrays a less fantastic sea monster than the text's neighing, snorting, horselike spouter that more closely resembles the* Carta Marina's *Prister near the Faroe Islands.*

RABELAIS'S PHYSETER

In Book 4 of *Gargantua and Pantagruel*, the gigantic Pantagruel and his companions are voyaging through the fabled Northwest Passage in quest of the Oracle of the Holy Bottle. Chapter 33, "How Pantagruel discovered a monstrous physeter, or whirlpool, near the Wild Island," evokes not only Olaus's horselike "Whirlpool," but also the *Carta Marina*'s adjacent Faroe Islands. Rabelais's tall tale parodies the traditional source of Olaus's story, Pliny.

About sunset, coming near the Wild Island, Pantagruel spied afar off a huge monstrous physeter (a sort of whale, which some call a whirlpool), that came right upon us, neighing, snorting, raised above the waves higher than our maintops, and spouting water all the way into the air before itself, like a large river falling from a mountain.

Pantagruel's fearful companion, Panurge, moans that the sea monster is the Leviathan of Job and pleads for some valiant Perseus to kill the beast. There is such a hero. Pantagruel showers the physeter with arrows until it rolls over on its back, "as all dead fishes do."

THE ZIPHIUS

Map D Ⓓ

A winding easterly course past the Faroe Islands approaches what Olaus and others call the Ziphius, a variant of *xiphias*, which derives from a Greek word for "sword." Olaus's commentary comes from a *History* chapter describing "the Sword-fish, Unicorn, and Saw-fish," sea beasts whose swords, horns, and saws bore holes in ships and drown mariners. Olaus omits the previously described Sea Unicorn here.

"Because this Beast is conversant in the Northern Waters, it is deservedly to be joined with other monstrous Creatures. The Sword-fish is like no other but in something it is like a Whale. He hath as ugly a head as an Owl: His mouth is wondrous deep, as a vast pit, whereby he terrifies and drives away those that look into it. His Eyes are horrible, his Back Wedge-fashion, or elevated like a sword; his Snout is pointed. These often enter upon the Northern Coasts, as Thieves, and hurtful Guests, that are always doing mischief to ships they meet, by boaring holes in them, and sinking them … The Sawfish is also a beast of the Sea; the body is huge great; the head hath a crest, and is hard, and dented like to a Saw: It will swim under ships, and cut them, that the Water may come in, and he may feed on the men when the ship is drown'd. There is also another sort of Saw-fishes that riseth against Marriners, that presently after 30 or 40 Furlongs is weary; and goes down into the Sea. The Seamen are often wounded with the sword of the Orca, which sticks upon his back, that they die of it: So by touching the Torpedo, their hand is drawn back stupefied."

The Orca/grampus/killer whale is the kind of monster that attacked the Balena and her calf near the island of Tile. A supreme marine predator, it preys on sea life from squid and seals to great whales. The "sword" on its back is similar to that of the owl-faced Ziphius, even though the two beasts are pictured and described differently. The Torpedo is a small electric ray.

The course skirts another of the map's three compass roses in the North Sea. Carta Marina *authority Edward Lynam reveals that these standard navigational elements, such as the map itself, are inclined to the east, not to true North.*

AMBRA
SPERMA
CETI
A
FAREN
STREME
ZIPHIVS
E
D
G
VACCA MARINA
HATKLYFTA

THE ZIPHIUS

Carta Marina

The story behind Olaus's sea beasts whose sharp weapons on backs and snouts can pierce the hulls of ships is yet another dreamlike blending of *Carta Marina* monsters with classical and medieval traditions. They morph from one animal into another under different names, such as the Orca and the Ziphius, the "Saw-fish" and the bestiary serra.

Ancestral Lore

Olaus himself cannot identify the smaller "grisly" beast, but the map legend describes the larger creature, the one with a raptor's face and curved beak, as a "Ziphius." That "Sword-fish" is not, however, our rapier-billed prize of deep-sea game fishermen. Olaus's commentary clearly indicates that what he considers a sword is the Ziphius's dorsal fin, such as that of the Orca.

For his reference to the *Xiphias gladius,* Pliny consults his favored fish authority, Trebius Niger, who stated that the swordfish pierces and sinks ships with its "pointed beak." Isidore of Seville, for one, repeats the detail, and Albertus Magnus, somewhat more cautiously, writes in his "Gladius" entry that the swordfish "reportedly impales the sides of boats." Albertus adds that he had once seen and examined the carcass of such a fish. Whether or not Olaus heard Scandinavian tales of fish destroying ships with their "swords," he localizes traditional knowledge, saying that such unwelcome guests invade northern coasts to do mischief.

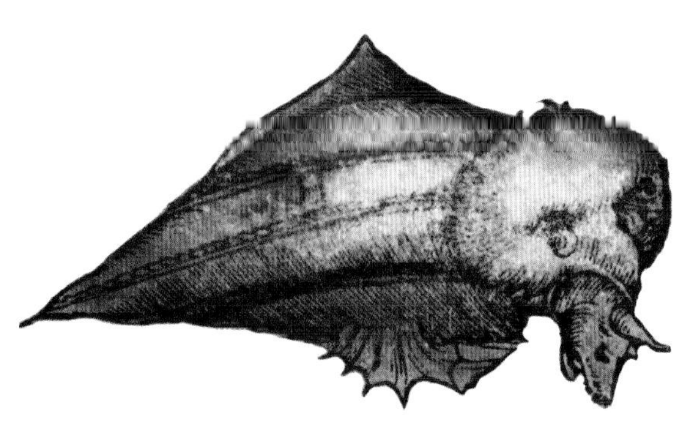

Sebastian Münster's ziphius, adapted from the Carta Marina *figure without the unnamed "grisly monster." The "seal" the ziphius is devouring is portrayed as a sea calf, a literal representation of that pinniped's common name.*

The owl-faced zifius is a quadruped in Hortus Sanitatis *("The Garden of Health," 1491). That standard reference work was the source of the Sifius in the later Dutch translation,* Der Dieren Palleys *("The Palace of Animals").*

A colored engraving of Conrad Gesner's ziphius. In the accompanying text, he credits Olaus with the image of the horrible sea monster.

Olaus begins his Ziphius commentary by adapting Albertus's "Xysyus" entry, which is perhaps the medieval scholar's second description of the swordfish. The Xysyus is a cetacean with a "deep, yawning mouth, awesome eyes, and a general body configuration unlike any other animal." Olaus's exaggerated version of the passage follows his comparison of the beast's ugly head to that of an owl. Pliny had said the fish had a "pointed beak," and Albertus refers to a swordlike snout.

Neither authority likened the fish's head to that of an owl, but *The Palace of Animals,* following the *Hortus Sanitatis,* does in its "Sifius" (swordfish) entry. And it even attributes the beast's distinguishing feature to Albertus. Beneath a portrait of the revered Doctor of the Church and the spiky owl-faced quadruped line, the brief text repeats Albertus's problematical Xysyus entry—which contains no owl comparison. The sifius came from other sources, and Olaus blended that minor tradition with Albertus's, just as *The Palace of Animals* did.

Map Legacy

The mixed scholarly and hearsay sources of Olaus's swordfish Ziphius did not influence the cartographers and naturalists who copied the vivid, dramatic figure. Variations of the Ziphius appear on the sea monster charts of Sebastian Münster (H) and Abraham Ortelius (E), in a Conrad Gesner natural history, and in Adriaen Coenen's personal manuscript. In all of them, the owl-faced beast is feeding on a seal—as orcas do—and all except *Islandia* present a swordlike fin on its back.

Both Münster and Ortelius echo Olaus's key entry (D d)—"The terrible sea-monster Ziphius devouring a seal"—in their own keys. Gesner, too, calls the Ziphius a horrible marine monster, and then cites Albertus. Coenen repeats Olaus's entire chapter on a page "as Olaus describes them." The cartographers do not mention the mixed ancestry of the Ziphius, but another Gesner woodcut depicts a long-snouted "swordfish."

And Since

In his *Pictorial History of Sea Monsters and Other Dangerous Marine Life* (1972), James B. Sweeney devotes much of a chapter to documenting what Pliny and other authorities said about the dangers swordfish (not orcas) pose to ships and sailors. Sweeney tells of sword-transfixed ship timbers in museum exhibits and relates nineteenth- and twentieth-century instances of swordfish attacking vessels from motorboats to schooners.

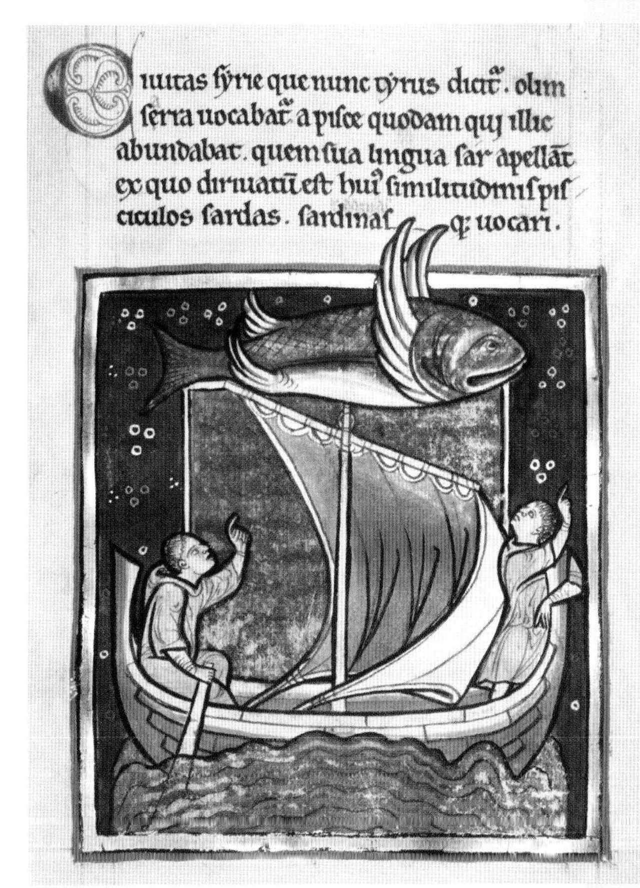

A twelfth-century bestiary illumination of a serra (sawfish) as a winged fish. From the Worksop Bestiary, England, possibly Lincoln or York.

THE BESTIARY SERRA

Olaus's Ziphius is a fantastic image. The sawfish in the *History* chapter's vignette of dangerous marine creatures is not. *Carta Marina* scholar John Granlund affirms that Olaus's sawfish is the first "realistic" picture of the fish with the distinctive serrated snout. Olaus emphasizes that the fish—like its sword-and-horned relatives—can destroy ships with its natural weapon.

He then adds that a different kind of sawfish swims with ships "30 or 40 Furlongs" until it tires and sinks back into the sea. That sawfish is the bestiary serra (from Latin, "saw, sawfish"). In the British Library's fourteenth-century Sloane 3544 manuscript, the sea monster is didactically explained as:

> *a symbol of those persons who at first early engage in good works, but who afterwards do not persevere in them, and are led astray by faults of different kinds, which toss them about as it were upon the waves of the sea and plunge them down to the depths of hell.*

The beast's serra/sawfish name notwithstanding, the bestiary animal could have been based on dolphins or flying fish. Medieval artists pictured it in any number of ways—from a lion-faced sea monster to a giant rooster.

THE SEA COW

Middle West E *(Not Keyed)*

A Sea Cow raises her horned head from the waters of the *Carta Marina.* Of the map's several sea monsters whose color, shape, size, or nature resembles that of a terrestrial animal, this figure is the most literal, as though it were a wandering farm cow. Olaus's description of the beast evokes lore of other marine creatures whose names identify their counterparts on land.

"The Sea-Cow is a huge Monster, strong, angry, and injurious; she brings forth a young one like to her self; yet not above two, but one often, which she loves very much, and leads it about carefully with her, whither soever she swims to Sea, or goes on Land; She is great ten moneths. Lastly, this Creature is known to have lived 130 years, by cutting off her Tail ... The Sea-horse, between Britany and Norway, is oft seen to have a head like a horse, and to neigh; but his feet and hoof are cloven like to a Cows; and he feeds both in the Sea, and on Land. He is seldome taken, though he grow to be as big as an Ox. He hath a forked Tail like a Fish. The Sea-Mouse makes a hole in the Earth, and lays her eggs there, and then covers them with Earth, on the 30 day she digs it open again, and brings her young to the Sea, first blind, and afterwards she comes to see. The Sea-Hare is found to be of divers kinds in the Ocean, but so soon as he is caught onley, because he is suspected to be venomous, how like soever he is to a Hare, he is let loose again. He hath four Fins behind his head; two whose motion is all the length of the fish, and they are long, like to a Hares ears; and two again, whose motion is from the back, to the depth of the fishes belly, wherewith he raiseth up the weight of his head. This Hare is formidable in the Sea; on Land he is found to be as timorous and fearful as a Hare."

Pictured with a Sea Cow in the accompanying vignette is a nursing sea calf that is similar to the Balena's suckling calf earlier in the voyage. The other animals in the vignette are described elsewhere in the *History.*

East of the Sea Cow on the large, fold-out Carta Marina, *the figure crossing a snowy mountain pass with his horse might be that of Olaus Magnus himself. And north of him are statues that ancient kings carved out of rocks to guide travelers.*

SPERMA CETI
ZIPHIVS
VACCA MARINA
GILLES
REDE
CASTRV
NIDROSIA METROPOL
SALTE
BACCA N MOSTE
VAROAL
AD KRAKAV
FOSE
BRVE
NERDE
HVMO DAL
KLEBO
LACVS ISTE NO CŌGELATVR
D KANVTVS ALSON
GISK
BESTAD
BEN
SALBO
SCORP E
BESTADA VDDEN
SKOGS BERG
VALHEM
MŌS ALTISSIMVS
HORNILLA BVK
STAD
HERŌ
OSTER
VOOS
HATKLYFTA
ASTRE
VALDRES
HALSNE
LADIA
MŌS
GIPSI
MVI
NORVEGIA
REGNVM

The Carta Marina *Sea Cow. Gerard Mercator reproduces the figure, and Conrad Gesner presents it, although he disapproves of heads of marine figures that are too similar to those of land animals.*

Olaus's Sea Cow joins the *Carta Marina*'s Sea Worm, the Sea Swine, and the Sea Unicorn in having a land animal's name. In his "Vast Ocean" commentary at the outset of this voyage, Olaus followed the long-lived tradition that all things on earth and in the heavens have their counterpart in the sea. That belief quietly lives on in our own marine-animal names, from catfish to starfish. How close the correspondences are is a question alluded to in classical times and debated during the New Science of the seventeenth century.

Ancestral Lore

Pliny the Elder established much of the land and sea lore—with subtle reservation—that Olaus inherited. Pliny refers to "the common opinion that everything born in any department of nature exists also in the sea, as well as a number of things never found elsewhere." He adds that the sea contains "likenesses of things and not of animals only," such as "the sword-fish, the saw-fish, and the cucumber-fish."

Ancestors of the fish composites that Olaus describes are mixed, as so many pre-Linnean lineages are. Olaus's Sea Cow (with the legend, "Vacca Marina"), comes from Albertus Magnus's "Vacca." Albertus translator James J. Scanlan equates that animal with the "Ox ray," whose fins resemble horns. That is not the "Sea Cow" pictured on the *Carta Marina* and in a *History* vignette. The "Sea Horse" might correspond to the walrus, whose Germanic name literally means, "horse-whale." The "Sea Mouse" can be traced through Albertus to Pliny to a filefish that digs a trench for its eggs, and the "Sea Hare" to a furry, poisonous fish in the Indian Ocean.

A sea cow, sea dog, and sea horse from the 1491 Hortus Sanitatis, *an herbal and natural history in the bestiary tradition.*

Map Legacy

Olaus's land/sea mutations aside, his realistic Sea Cow figure has its own influence on maps. Sebastian Münster's "monster" (T) is so-called because its head is similar to that of terrestrial cows (actually to that of the long-horned, now extinct, aurochs). Ortelius's engraver does not reproduce the *Carta Marina* Sea Cow, but substitutes for it what Ortelius calls "Seenaut" (K), gamboling sea oxen. Not far from those realistic land figures is the purely decorative hippocampus (G). Gerard Mercator pairs Olaus's bovine figure with the Prister (which was accompanied by the "Hippopotamus" legend) on his 1541 terrestrial globe.

And Since

The belief that terrestrial animals have counterparts in the sea is affirmed by Dithmar Blefkens in an account of his voyages and the history of Iceland and Greenland, *Islandia* (1607). No matter that he fabricated his 1563 voyage to Iceland and compiled his fiction from multiple sources, as John Mandeville did in his *Travels*. Iceland, Blefkens writes:

> *hath Horses and Kine, and what not: and it is a marvell how skilfull Nature sports,*

in expressing the shape of all earthly Creatures and Fowles in the Sea. Neither should any man perswade me that these things are true, although ten Aristotles should affirm them unto me, unlesse I had seene most of them with mine eyes. Let no man therefore presently cry out, that what he knows not is fabulous.

The land/sea analogy is one of the many "Vulgar Errors" that Sir Thomas Browne challenges in *Pseudodoxia Epidemica* (1646).

That all Animals of the Land, are in their kinde in the Sea, although received as a principle, is a tenent very questionable, and will admit of restraint. For some in the Sea are not to be matcht by any enquiry at Land, and hold those shapes which terrestrious formes approach not ... and some there are

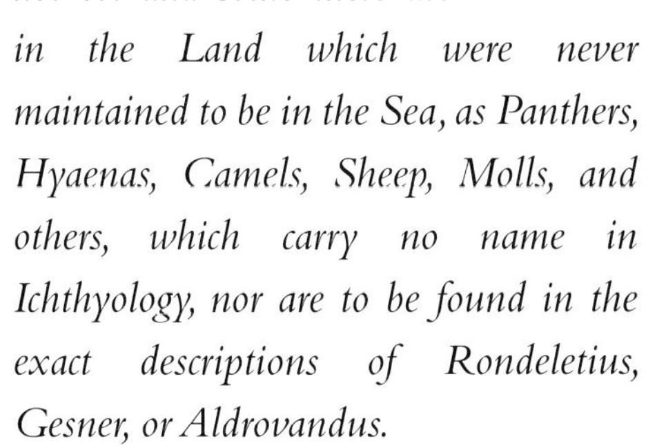

in the Land which were never maintained to be in the Sea, as Panthers, Hyaenas, Camels, Sheep, Molls, and others, which carry no name in Ichthyology, nor are to be found in the exact descriptions of Rondeletius, Gesner, or Aldrovandus.

It is fitting that Browne coined the word "ichthyology," for the study of marine life.

Steller's sea cow (Hydrodamalis gigas), *a large herbivorous relative of the manatee and dugong, hunted to extinction in the eighteenth century. An illustration from* Animals of the Prehistoric World *(1910).*

SEA/LAND ANIMALS

The Oxford English Dictionary identifies a host of marine animals with names of terrestrial counterparts. Common usage of most of the words ranges from the sixteenth through the nineteenth centuries.

Sea-bat—a flying fish

Sea-bear—the sea-urchin; the fur-seal; a polar bear

Sea-beaver—the sea-otter

Sea-calf—the common seal

Sea-cat—dog-fish (small shark); wolf-fish; sea-catfish

Sea-cow—the manatee and other sirenians; extinct Steller's sea cow; the morse or walrus

Sea-dog—the common or harbor seal

Sea-elephant—elephant seal, formerly the morse or walrus

Sea-hare—a mollusk with four tentacles

Sea-hawk—a flying fish

Sea-hedgehog—an echinus or sea-urchin

Sea-horse—the walrus; also the narwhal

Sea-lion—large-eared seals; a kind of lobster or crab

Sea-monk—the monk-fish (the angel-fish or the angler); the monk seal

Sea-mouse—a shell-fish

Sea-ox—the hippopotamus

Sea-wolf—wolf-fish; a seal; a sea-elephant or sea-lion

A Sea Rhinoceros

Map E (E)

Midway on the voyage up Olaus's map, a spotted creature with a swordlike hump and sharp curved nose is locked in battle with a giant lobster. The lobster is a recognizable animal. The larger, predatory beast that vaguely resembles a hippocampus seems to have emerged from a distant ocean. Olaus briefly describes the encounter, basing his commentary on only the *Carta Marina* key: "A monster looking like a rhinoceros devours a lobster which is 12 feet long."

What distinguishes Olaus's sea monsters from the legions of generic marine figures on antique maps—and makes them so influential—is that he cites them in his *Carta Marina* key and describes them in the map's commentary, his comprehensive *History*. Regardless of how fantastically the creatures are portrayed on the map, they are meant to represent actual marine animals. The rhinoceros-like creature is unique in that it is the only *Carta Marina* monster not acknowledged in the *History*. It is not even identified by a map legend. Because Olaus has so little to say about this particular beast, it is a most mysterious creature, making one wonder what it is and why it is in the map's northern waters.

Certain later critics would suggest a possible answer. A century after the printing of the *Carta Marina*, Sir Thomas Browne explains the cartographical use of sea horse figures. Hippocampi, he writes, are only "delineations which fill up empty spaces in Maps, and meer pictoriall inventions, not any Physicall shapes." And a century after Browne, Jonathan Swift's "On Poetry: A Rhapsody" (1733) contains the most frequently quoted lines about decorative animals on maps:

So Geographers in Afric-Maps
With Savage-Pictures fill their Gaps;
And o'er unhabitable downs
Place Elephants for want of Towns.

Olaus's rhinoceros-like monster does, indeed, decorate a blank space, but—similar to other *Carta Marina* monsters—it evokes the violence of animals in the perilous sea.

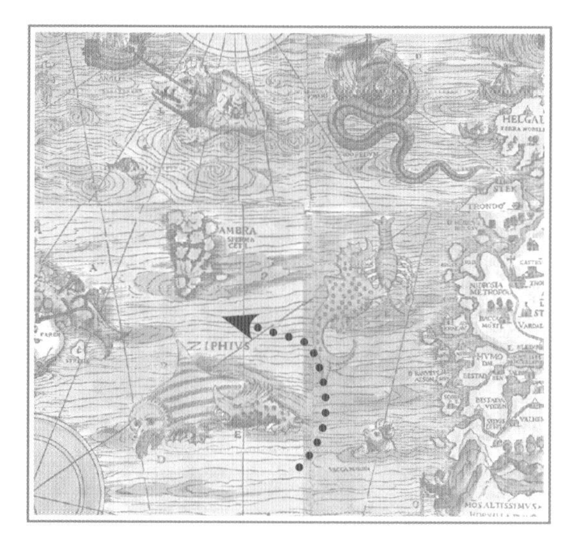

The voyage course bears westerly between the Ziphius and the fantastic beast gripping a giant lobster. East of the lobster's tail (E d) is a building with the legend of "Hericus Nielson." The full Carta Marina *key (see pages 150–151) indicates that, "Here they try to measure the unfathomable depths of the sea."*

HELGAL
TERRA NOBILI
OOPEDVM
STEK
TRONDO
AMBRA
SPERMA
CETI
D HERICVS
NIELSON
GILLES
REDE
CASTRV·ARE
TRONDI
NIDROSIA
METROPOL
SALTE
LADA
STRIN
BACCA
MOSTE
VAROAL
AD
KRAKAV
FOSE
ZIPHIVS
KLEBO
BRVE
NERDE
HVMO
DAL
LACVS. ISTE
NO CŌGELATVR
D KANVTVS
ALSON
GISK
BESTAD
BEN
SALBO
SCORP
E
BESTADA
VDDEN
SKOGS
BERG
VALHEM
VACCA MARINA
MOS ALTISSIMVS

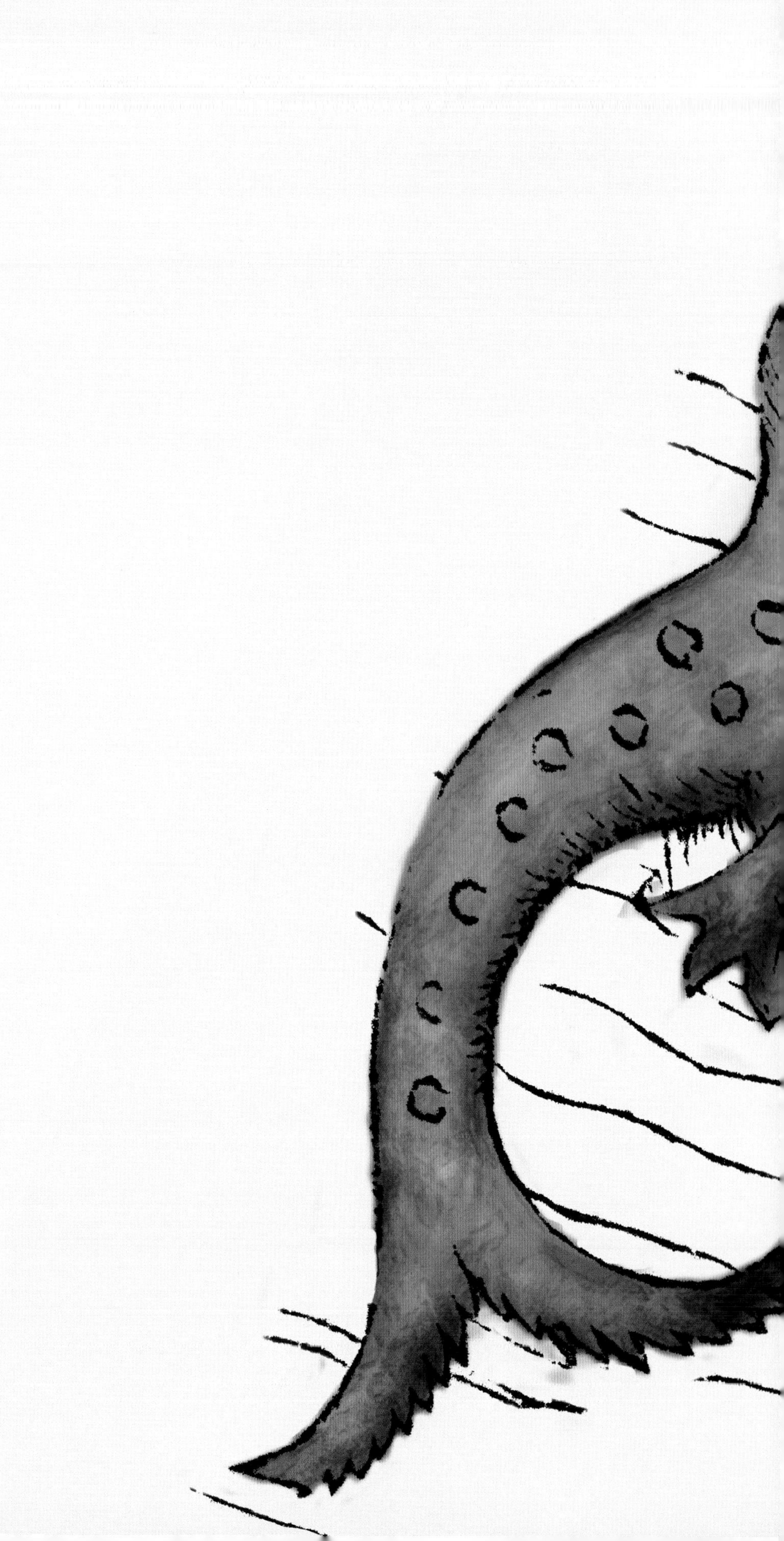

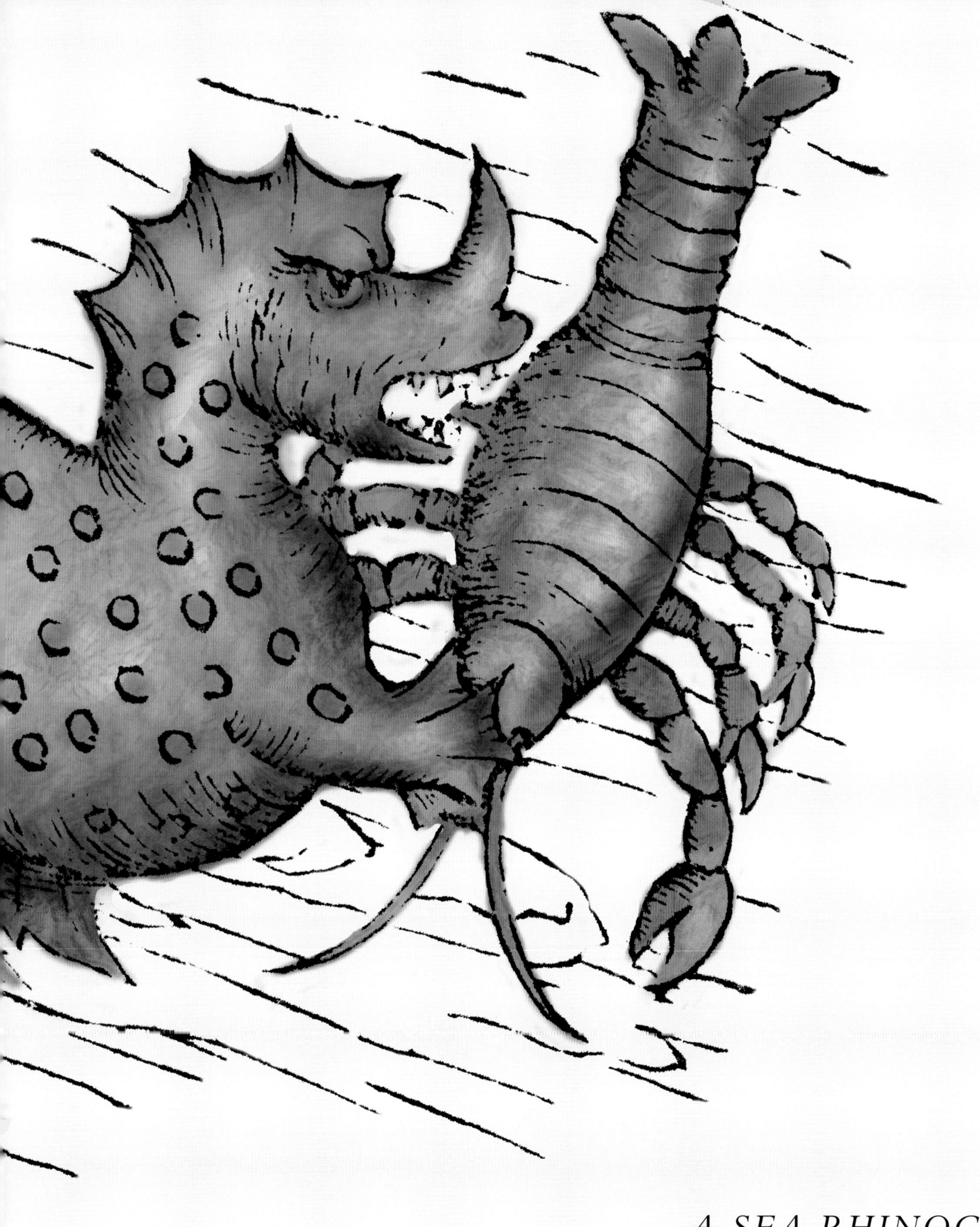

A SEA RHINOCEROS

Carta Marina

The *Carta Marina* key singles out the rhinoceros-like monster's similarity to one of the exotic animals reported by European explorers, made famous by Albrecht Dürer's celebrated woodcut, and pictured on early maps. Olaus's figure is one of his map's most bizarre images. The fantastic hybrid's fish tail, spotted body, flippers, dorsal fin, spiny crest, and horned nose are put together from the parts of various animals. Not included in the *History* and, thus, not placed within any written or oral tradition, the "monster looking like a rhinoceros" would seem to be Olaus's own invention of an exotic land animal's marine counterpart.

Its single horn places it in the age-old unicorn tradition, which is shared by the *Carta Marina*'s narwhal or Sea Unicorn.

Ancestral Lore

The rhinoceros had intertwined with complex unicorn lore since Ctesias's fifth-century BC account of the wild asses of India and the curative powers of their horn. Aristotle reinforced the tradition of a single-horned animal by declaring that there are two species: the Indian ass and the oryx. The first is thought to be the Indian rhinoceros, such as described by Ctesias, and the second one that had a lost a horn, or a profiled animal seen at a distance. Pliny recounts Roman games in which Indian rhinoceroses impaled elephants with their stone-sharpened horns.

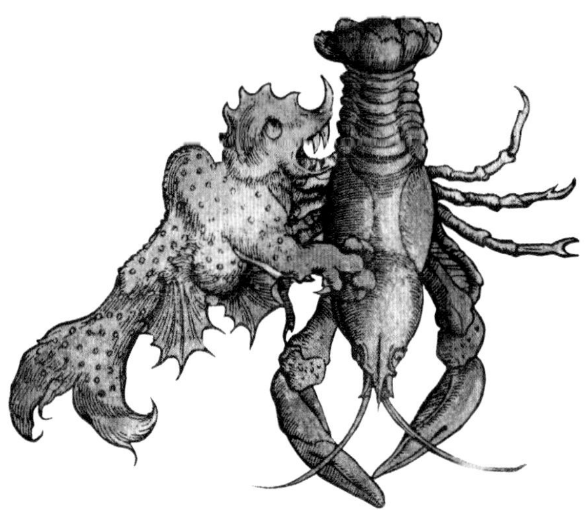

Sebastian Münster's speckled rhinoceros-like figure (N) adapted from the Carta Marina. *It, in turn, proved to be the model for Adriaen Coenen's "Rhinocer" on the facing page.*

Albrecht Dürer's famous woodcut of an Indian rhinoceros, based on a written description and a rough sketch. The image was reproduced in the works of Conrad Gesner and others. It influenced the depiction of exotic animals on maps and perhaps contributed to Olaus's Sea Rhinoceros.

The medieval conception of the unicorn as a mysterious equine figure with a long spiraled horn was well established by the time Marco Polo (ca. 1254-1324) describes an Asian rhinoceros in his *Travels*: "Tis a passing ugly beast to look upon, and is not in the least like that which our stories tell of as being caught in the lap of a virgin." In his *Allegory and the Migration of Symbols,* Rudolf Wittkower points out that the *Livre de Merveilles* (1403) illustrator of Marco's book depicted, instead, the graceful, white unicorns that the public expected—not rotund hairy animals that sleep in mud.

More than 1,000 years after amphitheater battles, the rhinoceros returned to Europe—and to lasting graphic fame. Albrecht Dürer's celebrated woodcut of a rhinoceros (1515) is the ultimate portrayal of an animal described. An inscription above the figure records that the King Manuel I of Portugal had received a similar beast from India on May 1, 1515. The text describes the animal's speckled coloring and hard scales and proceeds to quote Pliny on the enmity between rhinoceroses and elephants. Manuel I staged his own games, in which the elephant fled from the rhinoceros. En route to Rome as a gift to the Pope, the rhinoceros drowned in a shipwreck and its body was returned to Lisbon, where it was preserved by taxidermy.

While Dürer's armored beast is an imaginative work of art, not a totally accurate natural history depiction of an actual species, it did not take long for the figure to multiply on maps as well as on prints. Wilma George points out in her *Animals & Maps* of 1969 that cartographers began adding images of rhinoceroses to newly discovered lands in 1516. Most of those animals were based on Dürer's woodcut and sketches. The horned beasts join elephants, camels, and other exotic fauna in Ethiopia and India. These figures are not mere decoration to fill empty spaces but are serious attempts to illustrate the animals of distant countries.

Olaus was certainly aware of some form of Dürer's iconic image as its popularity spread throughout Europe. The unusual dotted circles on his *Carta Marina* figure would seem to imitate the "speckled" hide of the woodcut rhinoceros and its derivations on maps. In any case, Olaus chose to acknowledge the animal as a marine beast counterpart on his own map, citing it by name in his key and including its prominent feature—its horn—in the imaginary picture.

Map Legacy

The fantastic rhinoceros-like monster of the *Carta Marina* does not join the more naturalistic sea beasts on Ortelius's *Islandia*. However, although Olaus says so little about the creature gripping the great lobster, it is such a powerful image that Sebastian Münster, Conrad Gesner, and Adriaen Coenen adapt it in their works.

Adriaen Coenen's "Rhinocer," derived from Sebastian Münster's adaptation of Olaus's Sea Rhinoceros. Coenen's paintings of Olaus's sea monsters typically come from Conrad Gesner's natural histories or Münster's chart.

Münster repeats Olaus's key in his own key (N) with added details of the creature's sharp horn and pointed back. Nineteenth-century author John Ashton describes Gesner's version of Olaus's figure as a "Sea Rhinoceros or Narwhal"—a connection that Olaus himself does not make. Coenen copies Münster's version of Olaus's rhinoceros-like beast, which he describes as the "wondrous sea monster called Rhinocer."

Meanwhile, Albrecht Dürer's rhinoceros is disseminated through Münster's and Gesner's books and will be further popularized in Edward Topsell's natural history.

And Since

Olaus's narwhal and the rhinoceros-like monster, which appear opposite each other on the *Carta Marina*, have single horns that are part of the tangled myth of the unicorn. Sir Thomas Browne was one of the many scholars who joined the seventeenth-century debate over the nature of one-horned animals. Instead of denying that there was a unicorn, "we affirm there are many kinds therof." Those that he includes are: the Indian Ox, the Indian Ass, the Rhinoceros, the Oryx, fishes "described by Olaus, Albertus and others," and even four kinds of beetles.

SPERMACETI

Map D Ⓟ

A patch of what the *Carta Marina* legend identifies as "Ambra Spermaceti" floats upon the water. The emission of whales, it portends approach to the map's major whale grounds. The corresponding *History* vignette portrays sailors gathering the precious material for commercial use. Olaus describes the nature and medicinal benefits of the substance he himself has seen spread out upon the sea:

"The Whale copulating after the manner of men with the Female, by reason of the velocity of coition, he ejects much Sperm, and dissolves it; and when it is so dissolved, the Matrix doth not receive it all. It is scattered wide on the Sea, in divers figures, of a blew colour, but more tending to white; and these are glew'd together: and this is carefully collected by Marriners, as I observed, when in my Navigation I saw it scattered here and there: This they sell to Physitians, to purge it; and when it is purged, they call it Amber-Greece, and they use it against the Dropsie and Palsie, as a principal and most pretious unguent: It is white; and if it be found, that is of the colour of Gyp, it is the better. It is sophistiated [mixed] *with the powder of Lignum-Aloes, Styrax-Musk, and some other things. But this is discovered because that which is sophistiated will easily become soft as Wax, but pure Amber-greece will never melt so. It hath a corroborating force, and is good against swoundings and the Epilepsie."*

Olaus expresses the then common belief that spermaceti was the spilled reproductive sperm of a bull whale and that ambergris was processed from it. A confusion regarding the nature of spermaceti—a confusion that remains imbedded in our current word—began with the combination of Latin words for "sperm" and "whale." Spermaceti oil and ambergris were still regarded as a single whale emission a century after the printing of Olaus's map and *History*.

Northwest of sea beasts attacking each other, the course passes between a small floating island of "Ambra Spermaceti" and a captured whale that might have produced it.

ANGLI
GOTHI
L
AMBRA
SPERMA
CETI
A
P
FAREN
STREME
ZIPHIVS
E
D
VACCA MARINA

A drawing of a sperm whale (Physeter macrocephalus)*, showing the skeleton and the spermaceti organ, or case.*

Spermaceti is the oil within the enormous "spermaceti organ," or "case," of the sperm whale's head. It is in a liquid state at body temperature but solidifies when it is exposed to air. It is not whale sperm, as was widely thought for centuries. Nor does it become ambergris (French *ambre gris*, "gray amber," as distinguished from *ambre jaune*, resinous "yellow amber"). Found floating at sea or washed up on shore, ambergris is the intestinal secretion of whales. That Olaus, as well as others, regarded the two cetacean substances as only one is evident in the *Carta Marina*'s "Ambra Spermaceti" legend and the *History*'s chapter title. Like the transitional sea monsters, these two prized commercial products of the sperm whale evolve dramatically over time.

Ancestral Lore

Like many of Olaus's sea monster commentaries, his description of spermaceti relies heavily upon that of Albertus Magnus. The revered ecclesiastic and naturalist called discharged whale sperm ambergris, an effective treatment for various ills, including paralysis.

A contemporary of Albertus Magnus and compiler of the popular *On the Properties of Things,* the encyclopedist Bartholomaeus Anglicus also expresses the medieval view of ambergris as whale sperm/spermaceti: "It is said that the whale hath great plenty of sperm, and after that he gendereth, superfluity thereof fleeteth above the water; and if it be gathered and dried it turneth to the substance of amber."

In her *Monstrous Fishes and the Mead-Dark Sea*, Vicki Ellen Szabo notes that the lone medieval author who described ambergris and spermaceti as different substances extracted from different parts of the whale was Marco Polo (ca. 1254–1324). He relates in his *Travels* that on Socotra in the Indian Ocean, the islanders harpoon whales with barbed darts attached to long cords and buoys. After the marked whale dies and is towed ashore, "they extract the ambergris from the stomach and the oil from the head."

It is not until 1574 that a Flemish botanist, Carolus Clusius, became the first naturalist to identify ambergris as matter composed of squid beaks in the digestive tract of a whale.

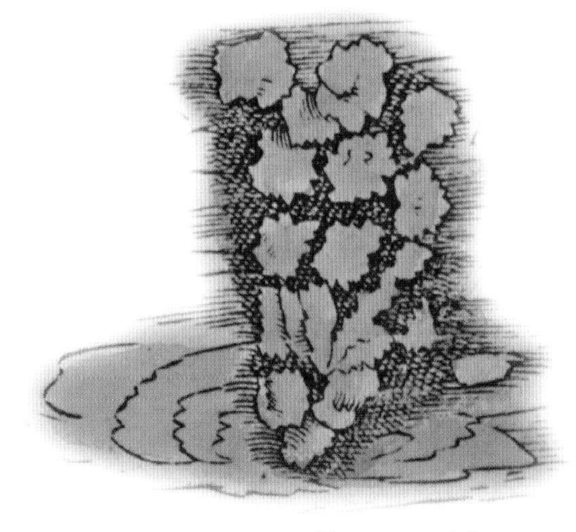

The spermaceti figure from the Carta Marina*, at the edge of the map's major whale waters.*

Map Legacy

Marco Polo and Carolus Clusius notwithstanding, the influence of Olaus's Spermaceti is evident on Abraham Ortelius's *Islandia* and in Adriaen Coenen's whale manuscript. Ortelius describes the two patches on his map as "Spermaceti, or a simple kind of amber, commonly called HUALAMBUR [whale amber]" (O). Coenen derives his ambergris-gathering scene from Olaus's *History* vignette and dramatically illustrates the supposed generation of spermaceti with a bull whale in hot pursuit of a female. The text of his "whale sperm" entry echoes Olaus's description of the substance and its harvesting, even to "(as I have seen at sea, writes Olaus)."

Engraving of a beached sperm whale. From the whale volume of The Naturalist's Library *(1833–1843), a forty-volume series edited by Sir William Jardine.*

And Since

It's not surprising that one of the "Vulgar Errors" of belief that the physician and celebrated stylist, Sir Thomas Browne, dissects in *Pseudodoxia Epidemica* (1672 edition) is conventional belief in the nature of spermaceti. He wastes no time in declaring that spermaceti "was not spawn of the Whale, according to vulgar conceit, or nominal appellation." However, he emphasizes that the substance comes from whales. Carolus Clusius and other learned naturalists established that, and proof of it was a spermaceti whale cast up on Browne's home coast of Norfolk in England. Browne describes in detail the streams of oil that flowed from the head of the whale, and he cites the oil's medical uses for treatment of "cuts, aches, and hard tumours." Browne clearly distinguishes between the spermaceti and the "Ambergreece," whose noisome stench prevented its extraction. After ambergris

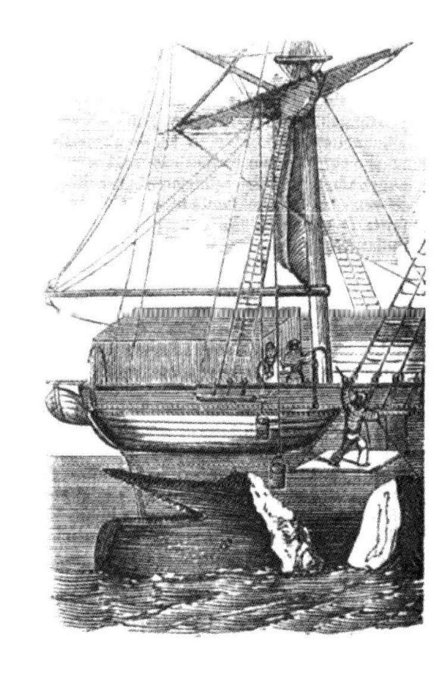

Removing spermaceti from a whale, from Robert Kemp Philp's Dictionary of Medical and Surgical Knowledge *(1864)*

was discovered in a sperm whale's stomach in the early eighteenth century, the odiferous substance also became highly valued—as a rare fixative of perfumes.

Browne and his contemporaries could not have envisioned that spermaceti—along with ambergris—would generate one of the greatest industries of the eighteenth and nineteenth centuries: sperm whaling. Already known for its medicinal properties, spermaceti oil replaced the oil of other whales as the fuel of choice. Bright-burning smokeless candles made from its wax were considered the best in the world. Its distilled oil filled the wells of lamps and lubricated machinery of the Industrial Revolution.

The golden age of nineteenth-century sperm whaling left in its wake the supreme literary evocation of the industry: Herman Melville's *Moby-Dick* (1851). Within its sublime pages is the passage in which the harpooneer Tashtego tumbles into the spermaceti case of a decapitated whale sinking beside the *Pequod*. The cannibal Queequeg dives into the fragrant morass to rescue the Indian. Soon after, "we saw an arm thrust upright from the blue waves; a sight strange to see, as an arm thrust forth from the grass over a grave." And the cry of "it is both!" greets the emergence of the two men. Second mate Stubb later greedily extracts ambergris ("worth a gold guinea an ounce to any druggist") from the bowels of a floating whale carcass.

A Beached Whale

Map D Ⓐ

Looming ahead beyond the floating Spermaceti are the Faroe Islands. On one, as the map key describes, "its fish-eating inhabitants cut up and divide among themselves the big sea animals thrown up by the storms." Midway up the *Carta Marina,* the voyage's progress slows before the charted course ventures into the most dangerous waters on the map. Olaus describes the beaching and the communal activity on the shore:

"Because the Whale falls greedily to eat the Herring, and Sea-Calves, as Fish fatter than all others; therefore is he often in danger on the Sandy Rocks, that by the ebbing and flowing back of the Waters are often left naked without Waters; & this Beast sunk into them, can find no way out that he might return to the next deep Water: and therefore he works so forcibly with his strong Tail, that he makes a large Ditch, and is bound up as in a Nest, the sand stopping him on all sides, that he cannot swim away. When Fisher-men know this, they run in Troops, and bind this Creature with strong Cords and Anchors, between his Chops and Gills, that he may not get off when the Sea come in; and with a strong hand, and many together, they draw him to the Land, or bind him so, that by no force he can be able to return to the Waters. And upon such a chance, they all joyfully divide the prey, and every one returns presently to his Houshold occasions, until the like, or more fortunate prize come again."

Olaus elsewhere recounts an ancient and recent stranding of whales, and the Scandinavians' many uses of the parts of the great animals: for food, clothing, heat and light, and even the very structure and walls of their homes.

The vignette that accompanies this commentary portrays not only the flensing of the whale, but also the village involvement in the dispersion of those parts in barrels and wagons.

At the southern approach to the Faroe Islands is a large rock aptly named the Monk. It is "an excellent protection against storms," as the moored ship indicates (D c on the complete key; see page 150).

NORDERO
DVMO
FARE
SVDERO
B
A
ECCLE
FAREN
MVLSE
STREME
C
MOACHV
AMBRA
SPERMA
CETI
ZIPHIVS
D
E

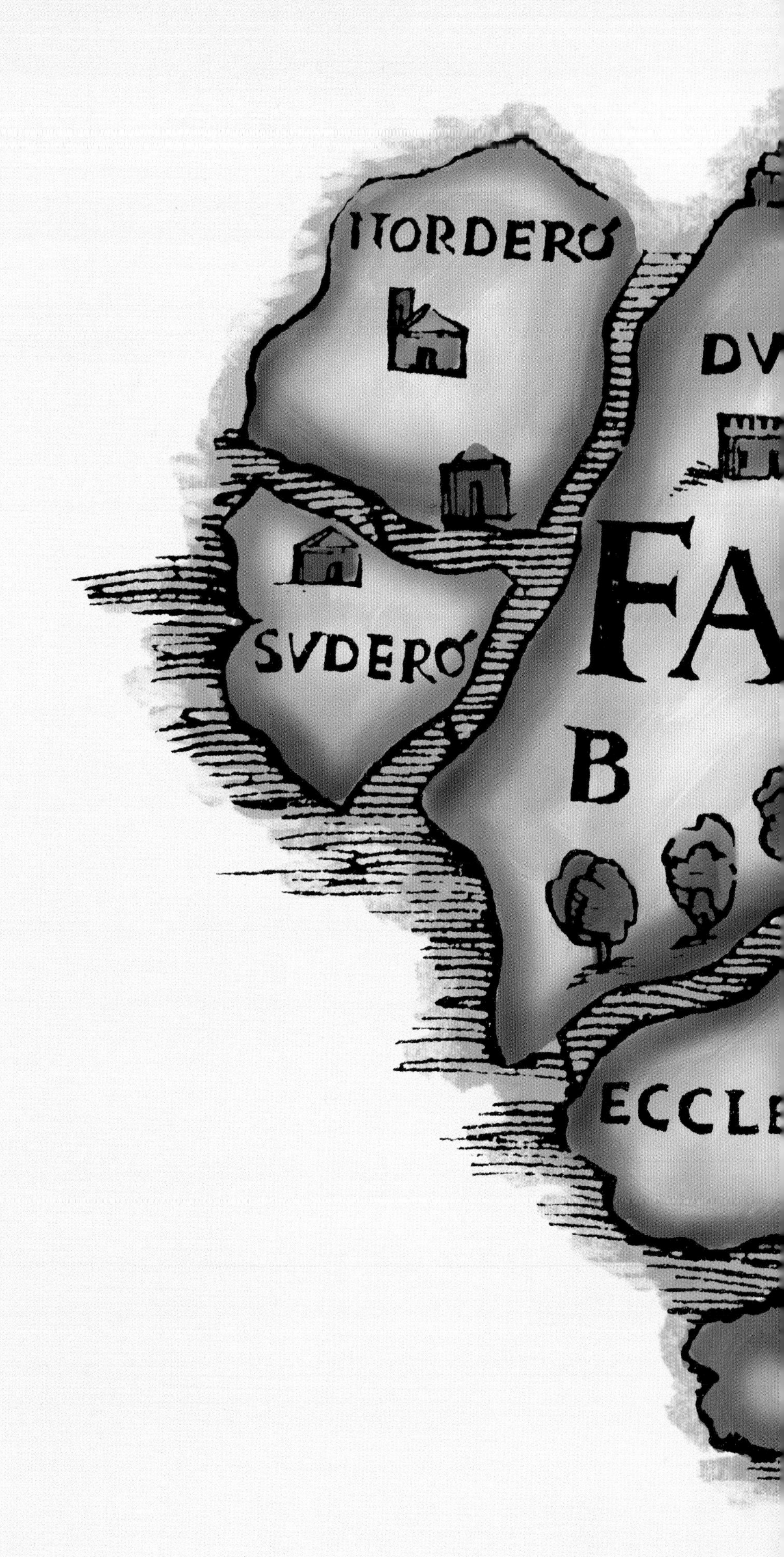

A BEACHED WHALE

Carta Marina

RE
A
FAREÑ

The Faroe Islands in the virtual center of the *Carta Marina*'s northern seas are a touchstone of reality halfway through the voyage. The map's waters in all directions teem with sea monsters born of mariners' tales, literary tradition, Olaus's observations, and an artist's imagination.

Between them, a bagpiper on a Faroe Island promontory heralds the stranding of a whale on the shore. An anchor is embedded in its back to prevent it from slipping into the sea, while ax-wielding men are attempting to hasten the whale's death. Because men are interacting with a sea beast on land, within community, the killing of the whale is similar to the everyday activities that fill the terrestrial portions of Olaus's map. The scene is the most naturalistic of any sea monster encounter on the *Carta Marina* and has its own legacy.

Olaus's History *vignette of the decaying Tynemouth whale, beached in northern England in 1532.*

Ancestral Lore

Olaus credits Procopius, a sixth-century AD Byzantine scholar, with an account of the stranding of a monster whale. Known as Porphyrios, this purple whale had menaced Constantinople and the surrounding area for fifty years, threatening seafarers and sinking ships with its weight. It was a time of earthquakes, the flooding of the Nile, and other disasters during the reign of Justinian when the beast, 45 feet long, became mired in mud at the mouth of Sakarya River. Villagers killed it and divided it among themselves. To some, its appearance was one more portent of doom.

Albertus Magnus examined the barrels of oil collected from a whale carcass he saw in the Netherlands district of Friesland. His extensive natural history entry on the whale details beach captures and the harpooning of whales at sea. Albertus's translator, James J. Scanlan, notes that these descriptions, based on personal observation and the reports of whalers, comprise "the first expansive treatment of the subject in Western literature." Olaus enhanced his borrowing of Albertus's writings with his own experience three centuries later. As Vicki Ellen Szabo summarizes in her *Monstrous Fishes and the Mead-Dark Sea*: "Olaus' greatest contribution to premodern maritime studies is found in his detailed look at the fishing industry of Norway, an account unparalleled by any other texts apart from his model and source, Albertus Magnus."

A whale stranding that Olaus does not picture on the *Carta Marina* but relates and illustrates in the *History* is that of "a monstrous Fish found on the Northern shore of England, Anno 1532." A "certain Noble-Englishman" recounted that by the time he observed the beast on the Tynemouth shore, the stench of

Conrad Gesner's naturalistic woodcut of the famed Tynemouth whale carcass. The bull whale's penis is extended when the muscles relax after death. Gesner does not cite Olaus as the source of the image.

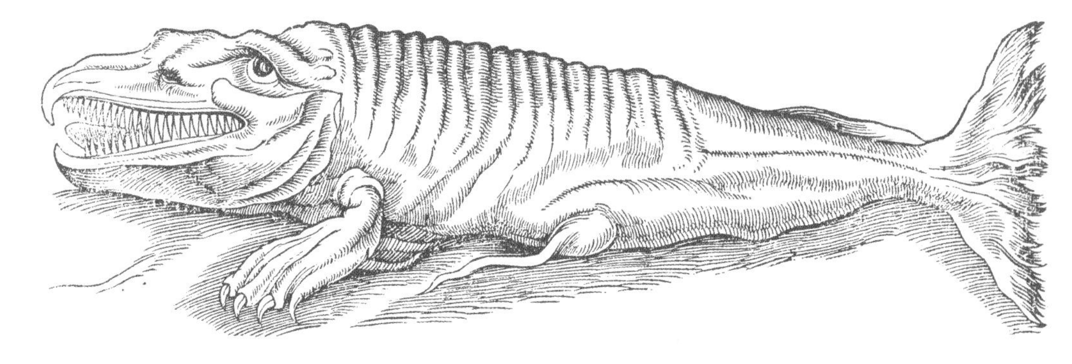

Conrad Gesner's elaborate rendering of Olaus's Carta Marina *figure of the beached whale on the Faroe Islands shore, with the bagpiper portrayed on the map moved to the head of Gesner's whale.*

the decaying carcass was overpowering. What still remained of the 90-foot monster would fill a hundred wagons. Although it was said the creature was not a whale because it had no teeth, its two blowholes and "Plates of Horn" indicate it was a baleen whale. Olaus adds that, "the Norway Coasts, between the mouths [fjords] of Berg and Nidrosum, have such a Beast as a constant Guest." Conrad Gesner's woodcut is similar to Olaus's vignette of the decomposing whale and was adapted by Adriaen Coenen and other naturalists into the eighteenth century.

Map Legacy

The most realistic whale on Olaus's chart has little place among the bizarre marine creatures depicted on Sebastian Münster's *Monstra Marina & Terrestria* and Abraham Ortelius's *Islandia*. Gesner's detailed rendering of Olaus's image appears on the same natural history page as the fanciful figure of the *Carta Marina* whale sinking a ship.

Olaus's and Gesner's beached whales are thus precursors of Hendrick Goltzius's 1598 engraving of a famous stranded whale at Katwijk, Holland. That lost drawing was said to establish such scenes as a conventional subject of art.

THE USES OF PARTS OF WHALES

Among the chapters in Book 21 of Olaus's *History* are a few that establish him as a major chronicler of Scandinavian maritime life as well as of sea monsters. Olaus explains how one beached whale can sustain a village:

> *"[O]f the flesh of one Whale, Fat & Bones, 250 or 300 Wagons may be loaded. They salt up the Flesh and Fat, in many great Vessels, amongst the rest of the huge Sea-fish; and they use these for their commodity, and Houshold-food, and they sell it to others for the same use, to be carried into remote parts of the World."*

The animal's skin is used to make belts, bags, bell ropes for churches, and enough clothing for forty men is obtained from a single whale. Whale oil burns in lamps, protects the hulls of ships, greases carriage wheels, and treats leather. Small bones provide household fires.

The great ribs, like an inverted keel, form the structure of houses where trees are scarce and bring the sea onto the land:

> *"They that sleep between these Ribs see no other Dreams than as if they were always toiling in the Sea-waves, or were in danger of Tempest to suffer shipwrack."*

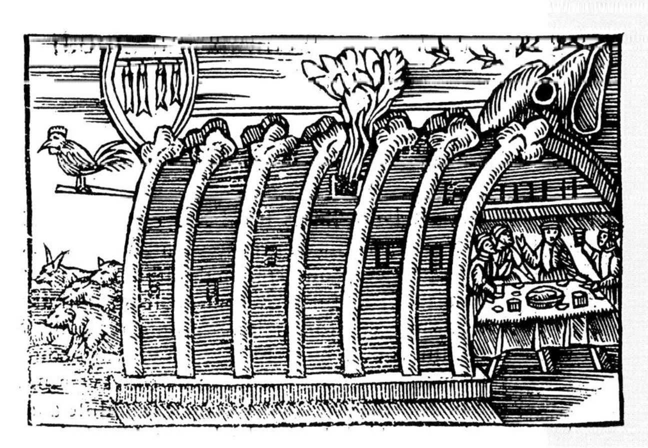

Olaus Magnus's History *vignette of a house constructed of whale ribs and hide. Dwellers gather for a meal inside the structure, which has windows, living quarters, and an area for animals. Outside is a perch for roosters to wake sleepers in the darkness of the northern winter.*

More Pristers

Map A Ⓚ

The voyage up the map becomes ever more perilous. North of the Faroe Islands, the waters begin to whirl in eddies. It is there that the warm Gulf Stream meets the cold Arctic waters south of Iceland. Within the swirling currents, as the map's key warns, "Sea monsters, huge as mountains, capsize the ships if they are not frightened away ..." Olaus explains how the seamen defend themselves from the threatening Pristers:

"The Whirlpool, or Prister, is of the kind of Whales, two hundred Cubits long, and is very cruel .... [A] Trumpet of War is the fit remedy against him, by reason of the sharp noise, which he cannot endure: and by casting out huge great Vessels, that hinders this Monsters passage, or for him to play with all; or with strong Canon and Guns, with the sound whereof he is more frighted than with a Stone or Iron Bullet; because this Ball loseth its force, being hindered by his Fat, or by the Water, or wounds but a little his most vast body, that hath a Rampart of mighty fat to defend it."

Olaus's commentary is from the *History*'s "Whirlpool, or Prister" chapter. The vignette above illustrates a different chapter in the complete *History*: "On the perils of sailors and the attacks of huge beasts, birds, and tiny fishes." Olaus again describes the driving off of assaulting whales—not only by means of empty barrels and trumpets, but also by pouring lye into the water. As the vignette portrays, the chapter also details the dangers mariners face from birds and from sucking fish. Sailors drive off flocks of attacking birds by singeing their wings with burning brands. The 6-inch remora (Latin "delay," or *echeneis*, Greek, "ship holder"), which Pliny called the greatest of nature's wonders, is more difficult to deter. It can stop a vessel under full sail by attaching itself to the hull, as it did ships of Antony at Actium and Caligula shortly before his assassination.

A team of oceanographers has speculated that the Carta Marina*'s swirls between Iceland and the Faroe Islands might be the first mapping of giant eddies in the northern seas.*

CHAOS
OSTRABORD
BREMEN
L
K
ANGLI
LUBICENSES
NORDERO
DUMO
FARE
SUDERO
A
B
ECCLE

MORE PRISTERS

Carta Marina

The first *Carta Marina* monsters to be seen in the turbulent waters north of the Faroe Islands are whales that a ship's crew attempts to drive off with trumpet blasts and bobbing barrels. The figure represents a pivotal stage in the evolution of a classical account into literal Renaissance illustration and then into proverbial usage.

Ancestral Lore

Tossed casks do not figure in a famous classical story of threatening whales, but trumpets are among the strategies of driving off the beasts. Olaus repeats Strabo's account of whales blocking the fleet of Alexander the Great as it sailed from India. In his *Geography*, Strabo (ca. 63 BC–AD 19) drew from the record of the fleet's commander, Nearchus, that what most distressed the men during the voyage was whales. The great beasts, "by their spoutings, would emit such massive streams of water and mist all at once that the sailors could not see a thing that lay before them." The pilots advised the terrified seamen that the sea creatures that caused the watery eruption could be driven off with trumpets and shouting. The fleet's loud charge into the pod dispersed the whales, which dived and breached astern the ships.

Pliny the Elder advanced an astronomical cause for the fleet's encounter with sea beasts. During solstices in the area of the Arabian Sea, he said, tempests raging down from the

Sebastian Münster's version of sailors attempting to drive off menacing whales (A). This ship is more crowded with passengers than its Carta Marina *model.*

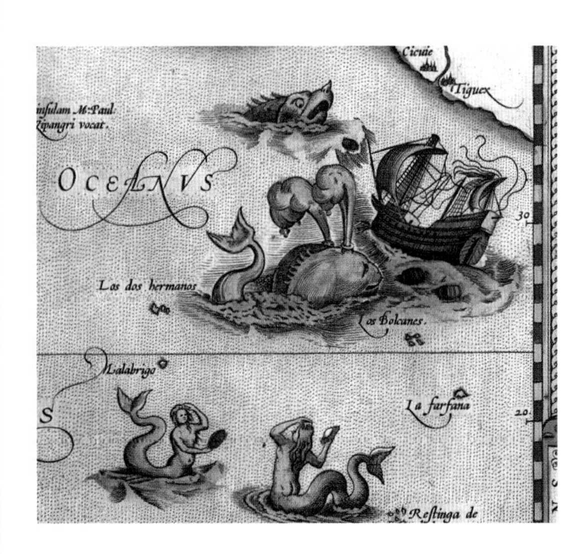

A threatening whale and barrels derived from the Carta Marina *join decorative mermaids and another sea monster on Abraham Ortelius's map of the Indies. The beleaguered ship is off the American coast of what is now California.*

THE ICELAND-FAROE FRONT

When oceanographer Thomas Rossby happened to see a reproduced portion of the *Carta Marina*'s northern seas in 2001, he was struck by the confluence of whorls in waters otherwise rendered in wavy lines. He immediately recognized that the map's spirals between the Faroe Islands and Iceland corresponded to satellite images of the Iceland-Faroe Front, an expanse of ocean eddies formed by colliding Gulf Stream and Arctic currents.

Rossby suspected that the map's eddies could be more than an artist's decoration. Although Olaus nowhere mentioned the curling lines, he seems to have added them to the chart to aid navigation. If so, Rossby concluded, they could be the earliest known representation of the Iceland-Faroe Front. Satellite images that were analyzed by his collaborator, Peter Miller, confirmed the theory.

The whorls on the *Carta Marina* spread across the most fearsome area of the map, where sea monsters attack ships and engorge sailors. It is tempting to think that Olaus placed these monsters where he did in order to warn mariners of dangerous waters.

Note: See Ocean Eddies in the 1539 Carta Marina by Olaus Magnus, *Oceanography, Vol. 16(4), 2003.*

mountains turn the sea upside down. Waves swirling up from the ocean bottom carry a multitude of monsters to the surface—like Olaus's Pristers rising out of the roily waters of the *Carta Marina*.

There are no war trumpets—or casks, or cannons—in Sinbad the Sailor's brief description of the dispelling of a giant fish. During his first voyage in the *Arabian Nights,* he comes upon fishermen in the Indies who drive away a sea beast 200 cubits long by beating sticks together.

Adriaen Coenen's watercolor based on Conrad Gesner's woodcut copy of Olaus's Carta Marina *figure.*

Map Legacy

Olaus's trumpeter clearly rises out of the classical tradition. The tossing of empty barrels overboard to distract the whales might simply be marine lore that Olaus gathered from Hanseatic League seamen.

The first reference to the maritime custom in English is thought to be in the anonymous *A Brief Collection ... gathered oute of the Cosmographye of Sebastian Munster* (1572). The passage, derived from the *Carta Marina* key, is elaborated only slightly from Münster's key (A). Great whales

> *will drowne and overthrowe shyps, except they be made a feared with the sound of trumpets, and drums, or except some round & empty barrels be caste unto them, wherewith they may play and sporte theym, because they are delited in playing with such thinges.*

An emblem adapted from the Carta Marina*'s spouting whales, floating casks, and trumpeter, from Joachim Camerarius's 1604 emblem book of fish and reptiles.*

Adriaen Coenen renders Olaus's trumpet and casks picture in his *Whale Book* manuscript, but his version of the tale differs from those of Olaus and Münster. The Dutch fish dealer/naturalist contends that the music of flutes and trumpets lures whales to the ships, where they are harpooned.

Variations of Olaus's iconic Prister image also appear on two maps later in the century, and in an early-seventeenth-century emblem book. Conrad Gesner's natural history copy of the *Carta Marina*'s menacing Pristers morphs into a figure on Christopher Saxton's 1583 wall map of England and Wales. Abraham Ortelius does not repeat Olaus's Pristers figure on *Islandia*, but his engraver depicts a similar scene on Ortelius's map of the Indies. The tableau is replicated in an emblem book of German scientist Joachim Camerarius the Younger.

And Since

The trumpet in Olaus's *Carta Marina* figure—and the iconic image itself—fade from reference as "to throw out a tub to a whale" gains status as a proverb meaning to distract an adversary. Jonathan Swift summarizes the nautical lore in his satirical *Tale of a Tub* (1704) and adds that his book is a metaphorical tub written to divert "the terrible Wits of our Age" from "sporting with the Commonwealth."

THE ISLAND WHALE

Map A Ⓛ

The whorls on the *Carta Marina*'s sea north of the Faroe Islands indicate strong swirling currents that threaten navigation. A snub-nose sea creature with long hair peers out of the roily water. Land rises dimly in the distance, offering mariners safe haven. Olaus explains that what might seem to be an island is not solid ground. His commentary derives from the *History* chapter, "Of Anchors fastned upon the Whales back."

"The Whale hath upon his Skin a Superficies, like the gravel that is by the Sea-side: so that oft-times, when he raiseth his back above the waters, Saylors take it to be nothing else but an Island and sayl unto it, and go down upon it, and they strike in piles unto it, and fasten them to their ships: they kindle fires to boyl their meat; until at length the Whale feeling the fire, dives down to the bottome; and such as are upon his back, unless they can save themselves by ropes thrown forth of the ship, are drown'd. This Whale, as I said before of the Whirlpool and Pristes, sometimes so belcheth out the waves he hath taken in that with a Cloud of Water, oft-times, he will drown the ship: and when a Tempest ariseth at Sea, he will rise above water, that he will sink the ships, during these Commotions and Tempests. Sometimes he brings up Sand on his back, upon which, when a Tempest comes, the Marriners are glad that they have found Land, cast anchor, and are secure on a false ground; and when as they kindle their fires, the Whale, so soon as he perceives it, he sinks down suddenly into the depth, and draws both men and ships after him, unless the Anchors break."

Olaus's matter-of-fact commentary, with its emphasis on anchors, would seem to be a treatise on Scandinavian whaling. It is actually an age-old story. A Catholic and a scholar, the nominal Archbishop of Uppsala paraphrases the standard Christian version from *Physiologus* and the bestiaries, but he does so without the overt religious moral.

The voyage course bears sharply to the east through the roily waters, away from the menacing whales and the bobbing casks.

ANGLI
L
L
GOTHI
OOPEDVM
AMBRA
SPERMA
CETI
A
E
FAREN

THE ISLAND WHALE

Carta Marina

The *Carta Marina*'s Island Whale is not the map's only whale with the fluke of an anchor embedded in its hide and men on its back. The captured whale on the Faroes shore is its counterpart. A major difference between them is that fishermen on the beached whale are employed in a realistic maritime activity, while the sailors cooking a meal on the Island Whale dramatize one of the oldest and most widespread legends of the sea. The two whales represent the mixed waters of the marine map.

In the Harley manuscript and other medieval bestiaries, sailors who mistake great whales for islands are like unbelievers who trust the devil and sink into hellfire. Harley MS 4751, f.69.

Ancestral Lore

Olaus's Island Whale has a longer and more extensive cultural lineage than any other marine figure on his chart. Sea monsters as big as mountains have appeared in stories worldwide since ancient times. One of these is Leviathan in the Book of Job, Psalms, and other Judeo-Christian writings. That immense beast is variously regarded as the sea itself, the king of all swimming creatures, the whale, and Satan. Given its distinction as the largest of all animals, the whale is definitely worthy of the leviathan name and of being the legendary fish of such great size that sailors mistake it for land.

An early version of the story is related at the beginning of Alexander the Great's *Letter to Aristotle* on the wonders of India (ca. AD 300, later in the medieval Alexander Romance). Alexander reports that natives told him of an island where a king was buried with much treasure. Alexander's close companion, Pheidon, insists on undertaking the dangerous voyage in his leader's stead. After Pheidon's crew lands, the supposed island suddenly "proved to be no island, but a monster which plunged into the sea." Bereaved Alexander later sees the beast, a formidable creature with tusks.

The seafaring tale takes on its standard form in the *Babylonian Talmud* (ca. AD 500), when the hero Rabha and his men land on what appears to be a beach of sand and grass. As Rabha tells it, "we thought it was an island, descended, baked, and cooked upon it. When the back of the fish grew hot, it turned over, and had the ship not been so near we would have been drowned."

Sinbad tells a similar, but more elaborate, story in his first *Arabian Nights* voyage. He and his shipmates are cooking, washing, and relaxing on a verdant shore when the sand begins to move and the captain on the anchored ship cries out to them to return to safety. The monster dives. Sinbad climbs onto a floating tub as the ship sails off, leaving him in the open sea. He drifts to an actual island and to new perils. The fact that the island-beast encounter is Sinbad's first adventure in a series of seven voyages and that it is the initial marvel of India recounted in Alexander's letter to Aristotle attests to the stature of the tale.

While he does not append a religious allegory to his whale with the anchor in its back, Olaus does elicit its Christian context. He cites the authority of fourth century Church Father St. Ambrose to confirm that whales and other sea monsters are so large that they resemble land. Ambrose writes in his *Hexameron* that whales are "of such huge bulk and measureless size" that "[i]f they were to float on the surface of the sea, you would imagine that they were islands or extremely high mountains whose peaks reach to the sky!" Ambrose was among the Fathers thought to contribute to the ancient *Physiologus*, which was Olaus's principal source for his seemingly realistic account of the Island Whale.

The massive *Physiologus* whale is called "Aspidoceleon" (asp turtle or shield turtle). Prefiguring Olaus's account, the sailors "plant their anchors" in the beast. What Olaus does not repeat is that the monster is a form of Satan and that, "if you fix and bind yourself to the hope of the devil, he will plunge you along with himself into hell-fire." The *Physiologus* adds that the demonic creature's sweet breath entices small fish into its mouth, just as the devil allures those of little faith.

The Old English *Physiologus* (tenth century) poetically extends the familiar details and Christian moral into an elaborate narrative about the fierce Great Whale. Although the monster is clearly a whale, its Anglo-Saxon name, "Fastitocalon," is the Greek-derived equivalent of the Latin-derived "shield turtle." The mariners "make fast their high-prowed ships by anchor-ropes to this land which is no land." After the men build a fire on its back, it plunges into the depths of the sea "and in that hall of death drowns sailors and ships." The Christian poem then vigorously expounds on the nefarious powers of the Whale as Satan.

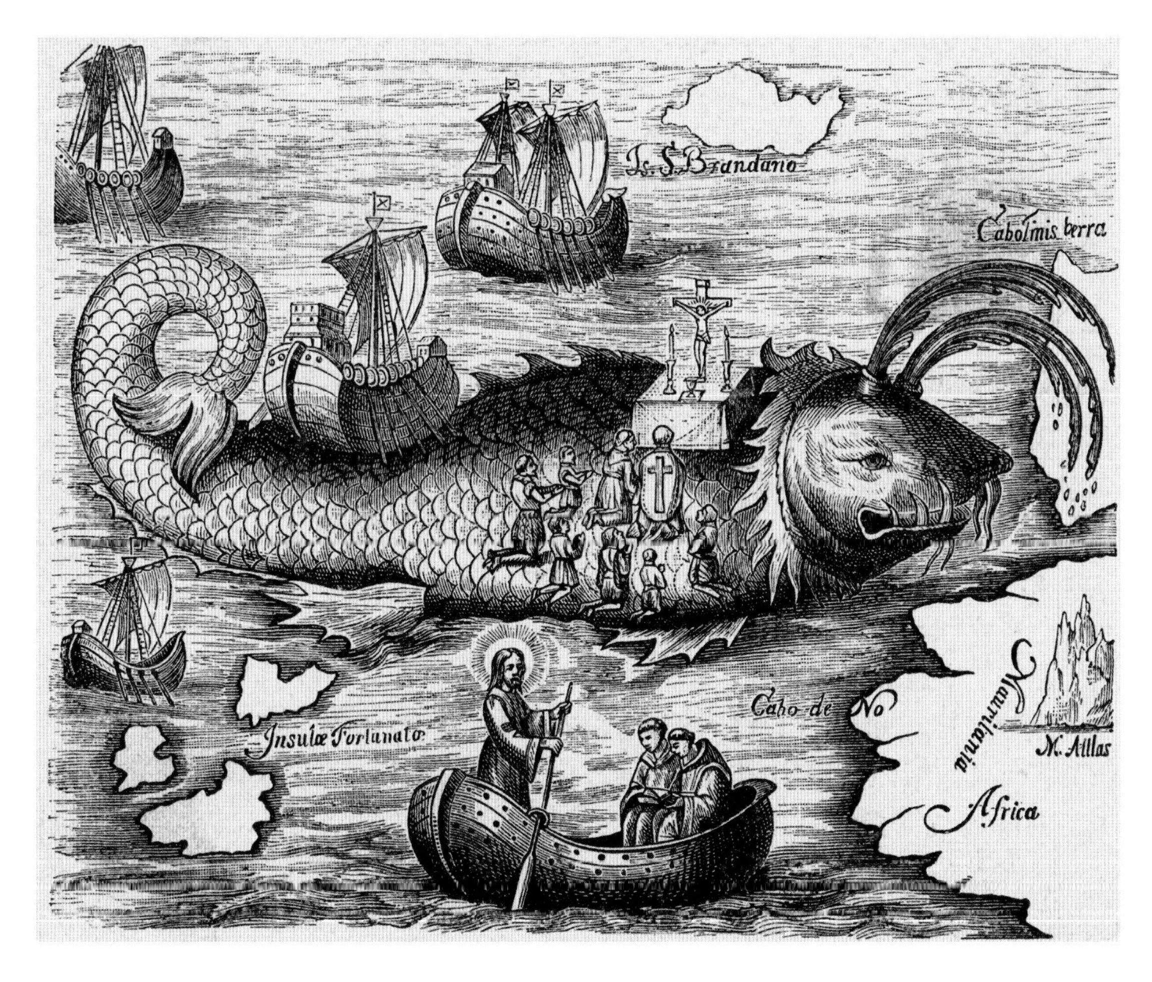

St. Brendan and his monks celebrate Easter mass on the back of the giant whale, Jasconius. St. Brendan's Island, the Promised Land of the Saints, is located to the north on this 1621 map by Honorius Philoponus.

Not all Christian versions of the legend are so grim. Olaus paraphrases the whale story of the Irish abbot, St. Brendan, who sailed the North Atlantic for years in search of the Earthly Paradise. The legendary abbot's encounter with Jasconius (the Latin form of the Irish word for "fish") differs considerably from the *Physiologus*/bestiary lesson. Brendan's monks celebrate Easter mass on what they (but not Brendan himself) believe to be a stony island. When the island begins to move, Brendan helps the brothers back into their boat and the fish dives.

The Brendan tale was frequently illustrated, and maps located the Earthly Paradise from Madeira to the West Indies for centuries thereafter. One of those maps is the 1621 chart of Honorius Philoponus, which places Jasconius between the Strait of Gibraltar and the Canary Islands, with St. Brendan's Isle to the north. The whale's inaccurate ruff, blowhole pipes, and watery spouting can be traced back to the whale figures on the *Carta Marina*. Similarities between the voyages of Sinbad and Brendan led to scholarly debate over which cycle of tales influenced the other, or whether they are unconnected and grew from independent sources.

Yet another approach to island-whale lore is Richard de Fournival's secular handling of the story in his thirteenth-century *Bestiary of Love and Response*. Fournival alters the standard religious moral by combining secular courtly love conventions with bestiary

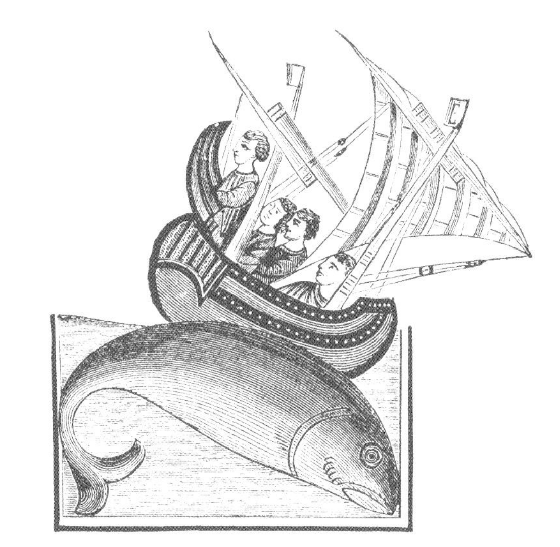

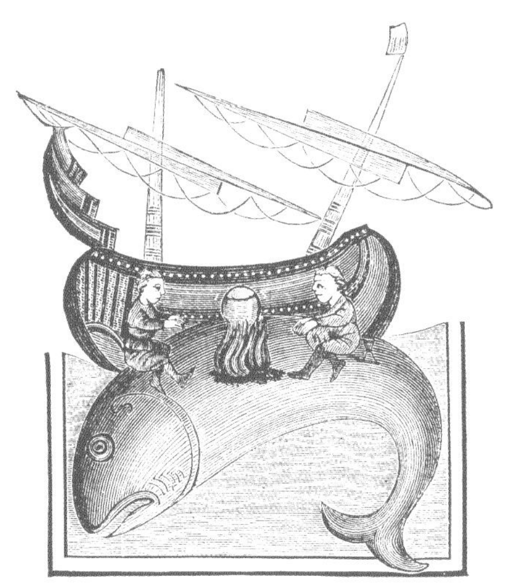

A tenth-century manuscript illumination of the standard Island Whale story, reproduced centuries later to illustrate Richard de Fournival's satirical Bestiary of Love.

traditions. "Master Richard" seeks the favor of his "fair, sweetest beloved" by sending her his bestiary. After relating the usual fable of the sailors cooking and the disturbed whale diving, he cautions her not be tricked by lovers less honorable than he is: "Wherefore I say that one must trust least whatever in the world appears most trustworthy. For this is what happens with most who become lovers. A man will say he is dying of love when he feels no pain or hurt, and these deceive good folk ..." She answers that the whale also dies.

Standard Olaus source Albertus Magnus says nothing about the tales of mariners boarding a whale, but he does dispel a belief about how the animal could become so large. It's said, he writes, that a bull whale becomes impotent after a single coupling and sinks to the sea floor, "where it grows and fattens to the size of an island. I do not believe this to be true."

Map Legacy

While Olaus's Island Whale derives from centuries of traditional lore, it nonetheless has its own small legacy.

Conrad Gesner includes in his *Historiae Animalium* a woodcut copied from the *Carta Marina* figure. Also, although the charts of neither Münster nor Ortelius depict an Island Whale with sailors cooking on its back, the keys to both charts reveal their indebtedness to Olaus. The first sentence of Münster's key (A) echoes Olaus's key and commentary, and

Ortelius acknowledges that the largest kind of whale resembles an island more than it does a fish (H). Amateur naturalist Adriaen Coenen attributes his illustration of the fabled scene to Olaus's account.

The most notable influence of Olaus's Island Whale is seen in John Milton's allusion to it in *Paradise Lost*. In Book I of the venerated English epic, Lucifer and his angels lie in the burning lake of Hell after their fall from Heaven. Satan lay "Prone on the flood, extended long and large," as great in size as

... that sea-beast
Leviathan, which God of all his works
Created hugest that swim the ocean stream:
Him haply slumbering on the Norway foam,
The pilot of some small night-foundered skiff,
Deeming some island, oft, as sea-men tell,
With fixed anchor in his scaly rind
Moors by his side under the lee, while night
Invests the sea, and wished morn delays:
So stretched out huge in length the Arch-Fiend lay
Chained on the burning lake.

Milton's references to both Norway and the "fixed anchor in his scaly rind" echo Olaus's Norwegian whale. Like *Physiologus* and the bestiaries, Milton here specifically compares the whale with Satan.

If nothing else, Conrad Gesner's often reproduced woodcut and the lines from *Paradise Lost* secure Olaus's Island Whale a small place in the great island-beast tradition.

The largest kind of whale, from Abraham Ortelius's Islandia *(H). The snout, teeth, spouts, ruff, and paws derive from* Carta Marina *whales.*

Adriaen Coenen's childlike drawing of the Island Whale is his own conception, unlike his other images derived from Olaus's map or History*. Coenen's manuscript page presents a pictorial narrative, from right to left.*

INSIDE LEVIATHAN

Closely related to stories of sea monsters so massive that they are mistaken for islands are tales of seafarers being swallowed by great beasts.

The thirteenth-century Bodley Bestiary combines the two archetypal motifs: "There is a monster in the sea which the Greeks call 'aspidochelon' ... it is also called sea-monster because its body is so huge. It was this creature that took up Jonah; its stomach was so great that it could be mistaken for hell ..." Jonah's plea for mercy is heard, and at the Lord's command, the whale vomits Jonah onto the shore. The scribe goes on to relate the island-beast legend and add its religious moral.

The satirical tall tale of the whale episode in Lucian's *A True Story* (second century AD) is entirely different. A whale 150 miles long swallows Lucian, companions, ship, and all. Inside the vast belly are a forest and an old man who had been there twenty-seven years, paying tribute to warring tribes. Lucian and his comrades rout the savages and set the forest afire. They celebrate their escape on the back of the dead whale.

And then there's the puppet Pinocchio and Monstro ...

THE SEA SERPENT

Map B Ⓓ

Ahead in the choppy waters of the *Carta Marina* is the most notorious and influential of all Olaus's marine monsters: the Great Norway Serpent, or Sea Orm. Born of mariners' tales, the terrifying creature that Olaus describes enters natural history and breeds serpent sightings across the Atlantic Ocean for centuries thereafter. He also tells of another such monster whose appearance was a portent of disaster.

"They who in Works of Navigation, on the Coasts of Norway, employ themselves in fishing or Merchandise, do all agree in this strange story, that there is a Serpent there which is of a vast magnitude, namely 200 foot long, and more—over 20 feet thick; and is wont to live in Rocks and Caves toward the Sea-coast about Berge: which will go alone from his holes in a clear night in Summer and devour Calves, Lambs, and Hogs, or else he goes into the Sea to feed on Polypus, Locusts, and all sorts of Sea-Crabs. He hath commonly hair hanging from his neck a Cubit long, and sharp Scales, and is black, and he hath flaming shining eyes. This Snake disquiets the Shippers, and he puts up his head on high like a pillar, and catcheth away men, and he devours them; and this hapneth not but it signifies some wonderful change of the Kingdom near at hand; namely that the Princes shall die, or be banished; or some Tumultuous Wars shall presently follow.

There is also another Serpent of an incredible magnitude in a Town called Moos, or the Diocess of Hammer; which, as a Comet portends a change in all the World, so that portends a change in the Kingdom of Norway, as it was seen, Anno 1522. That lifts himself high above the Waters and rouls himself round like a sphere. This Serpent was thought to be fifty Cubits long by conjecture, by sight afar off: there followed this the banishment of King Christiernus and a great persecution of the Bishops; and it shew'd also the destruction of the Countrey."

Fish with faces of men or lions were also considered portentous.

The dangers of the voyage intensify with the rising of the Norway Serpent. East of the monster is a perilous passage between it and the maelstrom whose "Caribdis" legend evokes Odysseus's threading between hideous Scylla and the great vortex.

LANGANES
ROST
XII·P.
HECEST HORRENDA
CARIBDIS
F
LOFOT
VAST
D
GOTHI
DLA
HELGALA
TERRA NOBILIVM
FISCA
STEK
TRONDO
D HERICVS
NIELSON
SPARABO
CASTRV·AREDI
GILLES
REDE
TRONDEM
NIDROSIA
METROPOL
SALTE
LESVNGER
LADA
STRINDAHERAD
BACCA
MOSTE
VAROAL
E
AD
KRAKAV
FOSE
YMBLOVIK
KLEBO
OFRESTRIND
D KANVTVS
BESTAD
SALBO

THE SEA SERPENT

Carta Marina

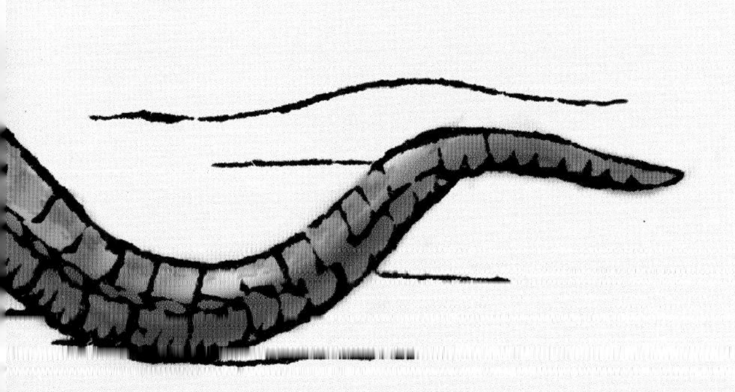

Olaus's commentary on his chart's Sea Serpent has the distinction of being the first written account of the great Sea Orm. Gleaned from the tales of Norwegian seamen, the marine monster has mythical and classical ancestors and a legion of descendants that have been pictured, analyzed, and debated up to our own time.

Ancestral Lore

A serpent myth with which Olaus and other Scandinavians would have been familiar is that of Jörmungandr, the Midgard Serpent from Norse mythology. After Odin cast the monster out of the home of the gods, it grew in the depths of the sea until it encircled the earth. Thor fished for it with ox-head bait, and hooked the beast, but the god's giant companion was so afraid the serpent would sink their boat that he cut the line. During the Twilight of the Gods, Thor and the serpent battled to their mutual deaths. The oral tales flourished in the Viking Age of sea dragon ships with serpent-head prows and curled-tail sterns. Olaus refers to those as "Dragons."

In natural history, Aristotle reported that an enormous Libyan serpent suspected of killing oxen upset a large trireme. Later in his Sea Serpent chapter, Olaus repeats another of the many classical accounts of serpents associated with water. According to Pliny the Elder, Roman general Marcus Regulus used weapons and catapults to destroy a 120-foot serpent at Bagrades River, in North

Africa. The remains of the creature, which is believed to have been a giant python, were displayed in Rome.

Elsewhere in his *Natural History*, Pliny mentions another representation of sea serpent lore, one the Roman public could view by the time Olaus himself visited and later lived in the city. It was the famous statue of the Trojan priest Laocoön and his sons writhing in the coils of a great snake sent from the sea as divine punishment. The Hellenistic sculpture was discovered in a vineyard on former imperial palace grounds and moved to the Vatican's Belvedere Garden in 1506.

Olaus does not mention the statue of Laocoön. As he says, his Sea Serpent is based on the accounts of Scandinavian mariners.

Map Legacy

A less terrifying version of Olaus's Sea Serpent wraps around a ship on Sebastian Münster's chart (C). Not surprisingly, the beast and sailors are nowhere to be seen on Abraham Ortelius's *Islandia*. However, the importance of the *Carta Marina* monster in early zoology is evident in the number of natural histories that reproduced the image. The most dramatic of all the map's marine figures, it has been the one most frequently copied or adapted by later artists.

The famous Conrad Gesner woodcut was originally accompanied by a depiction of the smaller, harmless sea snake that is harassed by a giant crab in the lower half of the *Carta*

Marina. In his 1608 *Historie of Serpents*, Edward Topsell copied both engravings along with commentary derived from Gesner. Topsell, however, erroneously describes the monster as being 120 feet long instead of the 100 to 200 feet that Gesner specified.

The sea serpent in Ulisse Aldrovandi's posthumous book of fishes was also obviously from Gesner, although it is presented more realistically, without ship or sailors. The engraving (labeled *Serpens Marinus Mari Noruegico familiaris Aldr.*) appeared in editions of John Jonston's *Theatrum universale omnium Animalium* into the eighteenth century.

Most sea monster studies dating from the nineteenth century—and now Internet Web sites—accompany discussion of the sea serpent with either the Gesner/Topsell woodcut or the vignette from Olaus's *History*.

And Since

Olaus's depiction and written account of the great monster are followed by centuries of sea serpent sightings and chronicles.

Scottish historian William Guthrie (1708–1770) reflects the serpent thinking of his time in his *Geographical, Historical, and Commercial Grammar; and Present State of the Several Kingdoms of the World* (published posthumously in multiple editions):

> *The fabulous sea-monsters of antiquity are all equaled, if not exceeded by the wonderful animals, which, according to some modern accounts, inhabit the Norwegian seas. Among these, the sea-snake, or serpent of the ocean, is one of the most remarkable, and perhaps the best attested.*

Facing page: The Norse Midgard Serpent lured to Thor's ox-head bait, from a seventeenth-century manuscript of the Icelandic School.

Sebastian Münster's "sea snake," up to 300 feet long. It harms sailors and sinks ships, "especially when it is calm."

A sea serpent derived from Olaus's Carta Marina *image appears on a 1982 South West Africa (now Namibia) stamp.*

While Guthrie does not cite Olaus by name, Dutch zoologist A. C. Oudemans is definitive about the marine monster's lineage. His monumental scholarly work, *The Great Sea Serpent* (1892), opens with references to Olaus in a chronology of "literature on the subject." The first ten of more than 300 items begins with Olaus's *History* and editions of the works of Gesner, Topsell, Aldrovandi, and Jonston that reproduce the serpent figure from the *Carta Marina*. These listings firmly establish Olaus's Sea Serpent as the major source of the great sea beast.

From the outset, Oudemans rejects Henry Lee's *Sea Monsters Unmasked* (1884) debunking of so-called sea serpents as gigantic calamari. Oudemans believes that Olaus and his artist intended to depict a serpent, although he considers the scales "badly drawn." He regards as fable Olaus's account of the Great Norway Serpent's nighttime forays to devour farm animals along the coast of Bergen. He also doubts that the creature could have snatched a sailor from the ship, because he knew of no such reported attack by a sea serpent. In spite of those objections, he accepts the beast as an actual animal and illustrates it with Gesner/

A nineteenth-century version of Parson Bing's drawing of the serpentine monster that Bishop Hans Egede describes in his A Description of Greenland *(1740).*

Topsell engravings and Olaus's vignette. Oudemans's seminal work documents the sightings that are recounted in virtually all studies of the subject thereafter.

Among standard sightings discussed in later books are early ones that were reported by eighteenth-century Scandinavian bishops, Hans Egede and Erich Pontoppidan, Bishop of Bergen, Norway.

Egede's eyewitness account is regarded as one of the earliest reliable descriptions of a "sea serpent," and Parson Bing's drawing as one of the first depictions of a sea monster based on a credible source. In *A Description of Greenland* (1740, 1818 edition), Egede reports that in 1734, he saw a "most dreadful monster" off the Greenland coast:

> *The monster was of so huge a size, that coming out the water its head reached as high as the mast-head; its body was as bulky as the ship, and three or four times as long. It had a long pointed snout, and spouted like a whale-fish; great broad paws, and the body seemed covered with shell-work, its skin very rugged and uneven. The under part of its body was shaped like an enormous huge serpent, and when it dived again under water, it plunged backwards into the sea and so raised its tail aloft, which seemed a whole ship's length distant from the bulkiest part of its body.*

Egede does not mention Olaus, but Pontoppidan does. Writing his *Natural History of Norway* (1755) during the Enlightenment, he concedes that Olaus "mixes truth and fable together according to the relations of others; but this was excusable in that dark age when that author wrote." Although Pontoppidan had questioned the existence of the sea serpent, "that suspicion was removed by full and sufficient evidence from creditable and experienced fishermen and sailors in Norway." Pontoppidan includes in his book a version of Parson Bing's dramatic rendering of the Egede serpent.

Authors have proposed multiple marine animals as those mistaken for the sea serpent: squid, whales, porpoises, sharks, oarfish, eels, pythons, and a host of others, as well as seaweed, driftwood, and fog. As early as 1822, Sir Walter Scott was one of the skeptics, notwithstanding his Romantic sensibility. The novelist opens Note 5 to *The Pirate* by conflating Scandinavian bishops while acknowledging that, "the wondrous tales told by Pontoppidan, the Archbishop of Upsal [Olaus Magnus—Ed.] still find believers in the Northern Archipelago." Scott cites "Guthrie's Grammar" as perpetuating that belief and goes on to relate an eyewitness account, not his own:

> *The Author heard a mariner of some reputation in his class vouch for having seen the celebrated sea-serpent. It appears, so far as could be guessed, to be about a hundred feet long, with the wild mane and fiery eyes which old writers ascribe to the monster; but it is not unlikely the spectator might in the doubtful light, be deceived by the appearance of a good Norway log floating on the waves.*

Olaus would certainly qualify as one of the "old writers" that Scott alludes to here.

Among the rash of nineteenth-century sightings in the Atlantic Ocean, one of the most authoritative—and, thus, one of the most sensational—was confirmed by officers of HMS *Daedalus*. The Plymouth, England, *Times* reported on October 7, 1848, that when the frigate "was on her passage home from the East Indies, between the Cape of Good Hope and St. Helena, her captain, and most of her officers and crew, at 4 o'clock one afternoon saw a sea serpent." The creature they described was about 60 feet in length and with head upraised some 4 feet above the water, propelled itself in a straight line at an estimated rate of fifteen miles an hour. London newspapers filled pages with accounts of the sighting and drawings of the described creature. The *Illustrated London News* feature on sea serpents included the Parson Bing drawing, a popular latter-day counterpart of the Gesner/Topsell engraving.

One of the most credible and sensational of all nineteenth-century sightings of sea serpents was that of the officers of the HMS Daedalus.

Sightings of sea serpents—and doubts of the monsters' existence—multiplied throughout the nineteenth century. The jury was still out when John Ashton concluded in his 1890 *Curious Creatures in Zoology* that, "I think the verdict may be given that its existence, although belonging to 'Curious Zoology,' is not impossible, and can hardly be branded as a falsehood." Belief in descendants of Olaus's Sea Serpent has not yet entirely been laid to rest, just as limited speculation as to the existence of "Nessie," the monster of Loch Ness, persists to this day.

Sketch of an oarfish that was considered to be a sea serpent found in Hungary Bay, Bermuda, January 22, 1860.

CARIBDIS

Map B Ⓕ

The voyage's charted course winds between the coiling Sea Serpent and the *Carta Marina*'s gyrating waters among the Lofoten Islands. In the center of the spiral is the speck of a doomed boat and crew. An accompanying legend identifies the foreboding figure: *"Hic est horrenda caribdis."* This is the horrible Charybdis. Olaus describes the care that mariners must take not to be sucked into the vortex.

"There are certain Bosoms of the Sea in my Gothick Map, or Description of the Northern Countries, that are engraved on the shore of Norway; namely Roest and Lo Hoeth; betwixt which, so great is the Gulf, that the Mariners that come thither unawares, are in a moment sucked in by its sudden circumvolution, all force and industry of the Pilot being taken away, especially those who know not the Nature of the place, or are otherwise driven on by the force of Tempests; or that, by contempt, little regard this imminent danger: Wherefore those that would sail thither from the Coasts of Germany hire the most experienced Marriners and Pilots, who have learned by long Experience, how by steering obliquely, and directing their course, they may avoid the danger better: and these are wont to bend their course over a great part of the Sea, by direction of a Compasse, that they may not fall into the Gulph; and chiefly about the most populous Cities Andanes, and Trondanes, and three other Islands, where part of the Sea is called Mosta Stroom; in which place the Flood is greater then in other waters about it. Also the Sea there, within the hollow Cave, is blown in when the Flood comes, and when it ebbs, it is blown out, with as great force as any Torrents or swift Floods are carryed. This Sea, as it is said, is sailed in with great danger, because such who sail in an ill time are suddenly sucked into the Whirl-pools that run round. The Remainders of Shipwracks are seldome restored again; and if they be restored again, they are so broken against the Rocks that they seem all in shivers and covered with hoariness."

Between the coiling Sea Serpent and the swirling maelstrom, fishermen remain close to the shore. Caribdis is swallowing the brown and white speck of a doomed craft. Carta Marina *authority John Granlund notes that Röst and Lofoten, islands adjoining the whirlpool, are misplaced in the vignette. They should be to the right of the vortex.*

C
CETE
GALLEA
PEREGRI
NA
OSTRAFIORD
EGGE
HASEVOG
LOKHI
GILLEFIORD
V
SANDER
LANGE
QVEDEFIORD
ROLLEN
LANGANES
TRODANES
XII·P·
ROST
DVVANES
ANDANES
SCONGEN
HECEST HORRENDA
CARIBDIS
F
D
LOFOT
VAST
GAMBLAVIK
PISCES
HVETE DICTI
NYGAVIK
DIA
HELGALA
TERRA NOBILIVM
HORV PISCIV
CAPITIBVS
VTVNTVR
LOCO LIGNO
FISCA
STEK
TRONDO
D HERICVS
NIELSON
TROLL
ARDAL
SPARABO
CASTRV AREDI

Olaus put the Lofoten Maelström (also "Mosta Stroom," "Moskestraumen") on the map, and his sensational description of the celebrated tidal current may in all likelihood be the first written account of it as well. His selection of "Caribdis" as the legend name for the *Carta Marina* figure evokes the whirlpool's classical antecedent, Charybdis—although he later says the Norwegian eddy is the stronger of the two. The dangerous Scandinavian wonder that Olaus introduced to the rest of Europe eventually spread throughout popular culture in nineteenth-century fiction.

Ancestral Lore

Of all the wanderings related about Homer's Odysseus, tradition has accepted the location of Charybdis (and Scylla) more than any of his other adventures. The Greek geographer Strabo (ca. 63 BC–AD 19) regards Charybdis as a myth, but he defends Homer as a geographer, contending that the poet derived his spouting and sucking sea monster from the ebb and flow of tides in the Strait of Messina, between Sicily and Italy. Currents of that strait pull ships bow-first into the "monstrous deep."

The hazard that is now described as the Scandinavian "Charybdis of the North" was *havsvelg* ("hole in the ocean") to the early Norse. The Nordic word "Maelström" is traced to the Dutch *malen* ("to grind") and *stroom* ("a stream").

BETWEEN SCYLLA AND CHARYBDIS

The witch Circe warned Odysseus that if the Sirens did not beguile him and his crew onto their treacherous rocks, he must sail between the six-headed Scylla and the engulfing whirlpool of Charybdis. Homer's wandering hero captivates the Phaeacian court with his account of that passage:

> *Then we entered the straits in great fear of mind, for on the one hand was Scylla, and on the other dread Charybdis kept sucking up the salt water. When she vomited it up, it was like the water in a cauldron when it is boiling over upon a great fire, and the spray reached the top of the rocks on either side. When she began to suck again, we could see the water all inside whirling round and round, and a frightening roar sound all round the rock. We could see the bottom of the whirlpool all black with sand and mud, and the men were at their wits ends for fear. While we were looking at this, and were expecting each moment to be our last, Scylla pounced down suddenly upon us and snatched up six of my best men.*

Once beyond the two monsters, the seafarers arrive at the sun god's island.

An eighteenth-century engraving of Odysseus and his crew threading the perilous passage between man-eating Scylla and swirling Charybdis.

Athanasius Kircher's diagram of a hypothetical underground river from the Norwegian Maelström to the Gulf of Bothnia. From his Mundus Subterraneus *(1664).*

Map Legacy

Homer's Charybdis is a sea monster. In spite of its mythical legend, Olaus's Caribdis is a tidal whirlpool off the northwest coast of Norway. Unlike *Carta Marina* sea beasts, the figure is not copied or varied on the Münster or Ortelius charts or the natural histories of Conrad Gesner and Adriaen Coenen.

Nonetheless, Olaus's Caribdis becomes a source for an underground river theory of German polymath Athanasius Kircher (1601/2–1680). In his *Mundus Subterraneus*, the Jesuit scholar propounds that the Norwegian Sea waters that are sucked down in the great Lofoten Maelström flow east through underground caverns to emerge beyond Sweden in the Gulf of Bothnia.

And Since

In his 1841 short story, "A Descent into the Maelström," Edgar Allan Poe popularizes the whirlpool's terrors that Olaus had chronicled four centuries earlier. Captain Nemo's *Nautilus* plunges into "the world's navel" at the end of Jules Verne's *Twenty Thousand Leagues Under the Sea*, and Herman Melville's Captain Ahab vows to chase Moby Dick "round Good Hope, and round the Horn, and round the Norway Maelstrom, and round perdition's flames before I give him up."

The fictionalized "hole in the ocean" opens yet again in the cataclysmic animated finale of the 2007 movie, *Pirates of the Caribbean: At World's End.*

A History *vignette depicting a stylized vortex. The vignette illustrates a chapter on the whirlpool and the dangers of ice. Olaus states that pieces of ice in a barrel will melt at the same time as the floes from which they were taken.*

Harry Clarke's 1919 illustration for Edgar Allan Poe's "A Descent into the Maelström."

A DESCENT INTO THE MAELSTRÖM

Early in his fantastic tale, Poe hearkens back to Athanasius Kircher and others who imagine that the Norwegian vortex is "an abyss penetrating the globe." A fishermen tells the narrator of his terrifying encounter with the whirlpool:

> *Never shall I forget the sensation of awe, horror, and admiration with which I gazed about me. The boat appeared to be hanging, as if by magic, midway down, upon the interior surface of a funnel vast in circumference, prodigious in depth, and whose perfectly smooth sides might have been mistaken for ebony, but for the bewildering rapidity with which they spun around, and for the gleaming and ghastly radiance they shot forth, as the rays of the full moon, from [a] circular rift amid clouds ... streamed in a flood of golden glory along the black walls, and far away down into the inmost recesses of the abyss.*

The Maelström whitened the fisherman's hair and spewed him back out.

Such exaggerated treatments of their Moskestraumen have irritated local inhabitants, who regard the eddy fondly. According to a Lofoten guidebook, villagers enjoyed watching the whirling currents: "Today they say, 'The Maelström, ah yes, that was our television when we were kids.' "

Another Prister

Map B ©

Respite from passing between the *Carta Marina*'s writhing Sea Serpent and engulfing Caribdis is short-lived. Beyond a large halibut, yet another spouting

Prister emerges from the map's waters. Pressing its great weight upon a ship's prow, it tips the vessel until it sinks with all its crew. Olaus continues his explication of pillarlike Pristers that were first encountered as the horse-head spouter west of the Faroe Islands:

"The Whirlpool, or Prister, is of the kind of Whales, two hundred Cubits long, and is very cruel ... This Beast hath ... a long and large round mouth, like a Lamprey, whereby he sucks in his meat or water, and by his weight cast upon the Fore or Hinder-Deck, he sinks and drowns a ship. Sometimes not content to do hurt by water onely ... he will cruelly overthrow the ship like any small Vessel, striking it with his back or tail. He hath a thick black Skin all his body over, long Fins like to broad feet, and forked Tail, 15 or 20 foot broad, wherewith he forcibly binds any parts of the ship, he twists it about."

Olaus's earlier Prister commentaries described not only the "Whirlpool" (D o), but also threatening whales being repelled with trumpets and empty casks (A k).

Sebastian Münster's sea beast chart and key include the trumpet and barrels tableau (A) and the upright, spouting monster (B), but not Olaus's Prister submerging a ship. Neither Olaus nor Münster's engraver seems aware that the figure of Olaus's fantastic "Whirlpool" looks nothing like the other whales, although it is one. Olaus's malevolent Pristers are not included on Abraham Ortelius's more naturalistic *Islandia,* which depicts several kinds of whales but no ships, sailors, or swimmers. When Ortelius varies Olaus's menacing Pristers on his map of the Indies, decorative mermaids primp themselves near the spouting whale.

To the northeast, a ship and sailing raft hug the shore beyond the deadly current of the Maelström. The course passes a halibut with both eyes on one side of its head and rounds a leviathan submerging a Danish ship.

A
GLACIAE
C
CETE
DANI
GALLEA
PEREGRI
NA
QVEDEFIC
LANGANES
ROST
XII.P.
HECEST HORRENDA
CARIBDIS
F
D
LOFOT
VAST
GOTHI
HELGATA
TERRA NOBILIVM

ANOTHER PRISTER

Carta Marina

The *Carta Marina* Prister drowning a ship with its weight and spouting is the last violent image on the voyage up the map. Most of the monsters encountered earlier are menacing or attacking ships and seafarers or each other. The gentle Balena is the first whale met on the charted course, the unique horselike Prister the second, but this Prister is the last of its diverse cetacean family to show itself in the map's major whale grounds north of the Faroe Islands. The only *Carta Marina* whale shown physically sinking a ship, it presages nineteenth-century sperm-whale destruction of ships immortalized in the actual *Essex* whale ship disaster and the novel it inspired, Herman Melville's *Moby-Dick*.

Ancestral Lore

Olaus's final Prister sinking a vessel is even more aggressive than the pod of whales that Alexander the Great's commander, Nearchus, famously repulsed with the blasts of war trumpets and his fleet's military charge into their midst. Olaus repeats Strabo's story and Pliny the Elder's retelling. Roman historian Arrian (AD 86–160) also narrated the encounter in his *Indika*, adding that the sailors were so terrified by the whirlwind spouting of the whales that the oars dropped from their hands. In those classical accounts, no beasts actually ram or overturn the ships.

The whales that bedeviled Alexander's fleet are usually identified as sperm whales, the great toothed species associated with the "physeter" (Prister), and later with the 1820 *Essex* disaster. The thirteenth-century Old Norwegian manuscript, *The King's Mirror,* bypasses tradition by deeming the sperm whale a gentle beast that stays away from fishermen. The kind of whale most similar to Olaus's attacking Prister is the anonymous author's "horse whale" or "red comb." Such creatures "are very voracious and malicious and never grow tired of slaying men. They roam about in all the seas looking for ships, and when they find one they leap up, for in that way they are able to sink and destroy it the more quickly." The large humpback whale, too, is dangerous. It obstructs the course of ships, and when vessels sail across it, they capsize, drowning all. Other kinds of whales are helpful to fishermen, namely the "fish-driver," which herds schools of herring into the shallows, where they are easy to net.

A marine creature that could have served as a model for the figure of Olaus's final Prister is the similar Pistris. The late-medieval herbal, *Hortus Sanitatis,* pictures the

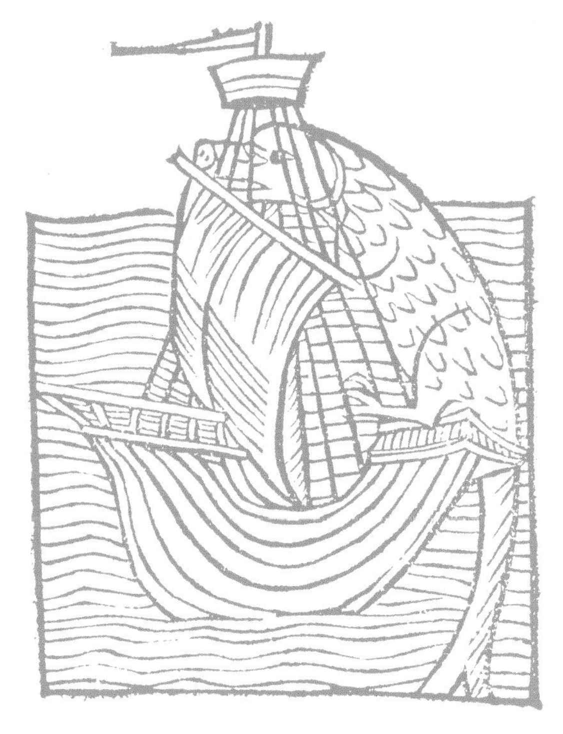

A Hortus Sanitatis *woodcut of a Pistris pulling itself onto a ship. The Pistris is either a Prister/Physeter or a close relative.*

Adriaen Coenen's panel of the Physeter, Pistris, and the Tinnus (left to right) assaulting ships. The Tinnus gives birth in the sea and eats on land. It has boarlike bristles on its back. Figures of the Pistris and the Tinnus derived from Hortus Sanitatis.

Pistris as a long-tailed, legged beast clambering onto a ship. As large as the vessel itself, it perches on the stern, its head among the shrouds above the mainsail.

Map Legacy

No variation of the *Carta Marina*'s ship-tipping Prister appears on either Sebastian Münster's *Monstra Marina & Terrestria* or Abraham Ortelius's *Islandia*, but two of Ortelius's whales and another of his monsters decorate Jan van Doetecum's 1594 map of the North Atlantic.

A woodcut copy of Olaus's figure notably opens Conrad Gesner's *Historiae Animalium* pages on Olaus's whales. Also, Adriaen Coenen groups the image along with three others on his "Physeter" pages: the pillarlike, horse-head Prister; the ship-climbing Pistris copied from the Dutch translation of *Hortus Sanitatis*; and Tinnus, a sea monster similar to the Pistris, also pulling itself onto a vessel. Coenen's accompanying text is from Olaus, without attribution.

And Since

The Prister pressing down the prow of a vessel is mild compared to nineteenth-century reports of sperm whales' provoked attacks on whale ships. The most famous such incident is the 1820 sinking of the *Essex*, which prefigures the White Whale's vengeful annihilation of Captain Ahab's *Pequod* in the timeless *Moby-Dick*.

The most famous actual incident of a bull whale attacking a ship is that of the ramming and sinking of the Essex, *in 1820. Another whale ship, the* Ann Alexander, *suffered a similar fate three decades later in 1851, the year Herman Melville's* Moby-Dick *was published.*

The climactic rising of Moby-Dick, the White Whale, in Herman Melville's classic 1851 novel. Illustration by A. Burnham Shute.

SINKING OF THE *ESSEX*

Herman Melville's *Moby-Dick* (1851) would not be the novel it is without the *Essex* tragedy and Owen Chase's 1821 *Narrative of the Most Extraordinary and Distressing Shipwreck of the Whale-ship Essex*. Chase's son gave Melville a copy of the first-hand account. In it, first mate Chase vividly describes the November 20, 1820, attack that leaves the ship's crew in open boats, thousands of miles west of South America.

Chase was onboard the *Essex*, repairing a whale-boat stove by one of the whales being hunted, when a large sperm whale surfaces, facing the ship. The whale spouts, disappears, and rises again. "He came down upon us with full speed, and struck the ship with his head, just forward of the fore-chains."

The whale swims off, and while Owen orders the starting of pumps and lowering of boats, he hears, "here he is—he is making for us again." Charging with "tenfold fury and vengeance in his aspect," the whale "completely stove in her bows. He passed under the ship again, went off to leeward, and we saw no more of him."

Eight members of the crew of twenty-one survived the three-month Pacific voyage.

A SEA CREATURE

Northwest B, Not Keyed

A northeasterly course skirts a mass of Icelandic ice where polar bears feast on fish. Below Greenland, a Sea Creature with a spiky, icicle-like beard surfaces. It looks like an old man of the sea, a male counterpart of the somewhat feminine creature near the Island Whale. Drawing on learned tradition and what Scandinavian seamen had told him, Olaus explains why marine beings with human features pose such a great danger to seamen:

"There are ... Monsters in the Sea almost like to men, that sing mournfully as the Sea Nymphs: Also Sea-men, that have a full likeness of body, and these in the night will seem to go up into the ships; and it is proved, that where they approach there is some danger coming; and if they stay long the ships are drowned. Also I shall add from the assertion of the faithful Fishers of Norway, that if they take such, and do not presently let them go such a cruel Tempest will arise, and such a horrid lamentation of that sort of men comes with it, and of some other Monsters joining with them, that you would think the Skie would fall; and the Fisher-men, with all their labour, can scarce save their own lives, much lesse can they catch fish. Wherefore in such a case, it is provided and observed by the Law of Fishing, that when such monstrous fish are drawn up as their form is various, so they must be presently let go, cutting away the Hooks and Lines."

The bearded Sea Creature rises in the far northern waters near "Islandia" and one of the map's sections of Greenland. Above the "Mare Glacia," on the large, fold-out *Carta Marina*, is the figure of a seated fiddler using a hunter's ploy of luring swans with sweet music. Spread along Greenland's storm-tossed southern coast is driftwood, which can penetrate hulls and sink ships. *Carta Marina* scholar Edward Lynam ventured that the rounded boat might be the first illustration of a type of skin kayak. He points out that the archer shooting at a ship is an Eskimo and that the triangular constructs on the shore are igloos.

Above the creature in the Greenland Sea are what the map key calls skillful sailors who use their hardy leather boats to attack the ships of foreigners (B a).

PARS
A
B
C
CETE
DANI
GALLEA
PEREGRI
NA
OSTRAFIORD
QUEDEFIORD
LANGANES
TRODANES
ROST
DUVANES
XII-P.
HECEST HORRENDA

The monsters that sing like "Sea Nymphs" and resemble "Seamen" join the *Carta Marina's* Sea Swine, Sea Cow, and others in the venerable tradition that all land animals have counterparts in the sea. Of all such creatures, those with human characteristics are the most ancient and universal in myth and folklore, beginning with the Babylonian fish-men. Olaus Magnus's Sea Creatures are members of that vast marine family that includes classical nereids and tritons, sirens, mermaids, and mermen. While fabled sea creatures that resemble comely young women entice sailors to their deaths, Olaus's Sea Creatures with human shapes can submerge a craft as surely as the *Carta Marina* Prister does.

An illumination from a late thirteenth-century Latin Physiologus, *from the Sloane manuscript. The sweet and deadly song of the Sirens lures one sailor to his death while another seaman covers his ears. The* Physiologus *lesson is that men are thus deceived by the charms of this world.*

Ancestral Lore

The beginning of Olaus's commentary is highly indebted to his principal classical authority, Pliny the Elder. Mythical sea-nymph daughters of Nereus and young male Tritons become creatures that the Roman naturalist famously describes. He relates that an emissary from Lisbon reported to the emperor Tiberius "that a Triton had been seen and heard playing on a shell in a certain cave, and that he had the well-known shape. The description of the Nereids also is not incorrect, except their body is bristling with hair ..." The song of one of these that was seen dying on the shore was heard far down the coast. Informants that Pliny trusted told him they had seen in the Gulf of Cádiz a human-formed creature such as Olaus describes in his commentary. That monster climbs onto ships at night, "and the side of the vessel that he sits on is at once weighed down, and if he stays there longer actually goes below the water."

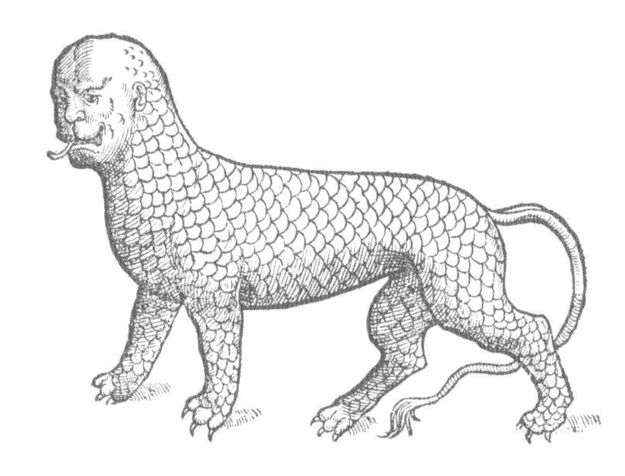

Conrad Gesner's lionlike sea monster with the face and voice of a man. The scaly beast was captured shortly before the death in 1549 of Pope Paul III and was exhibited in Rome.

Another of Olaus's frequent sources, thirteenth-century Albertus Magnus, cites Pliny's Nereids in his natural history. He states without further comment that they are "sea creatures which in some respects simulate a human form" and wail mournfully when they die. He classifies Homer's sweet-singing, deadly Sirens and six-necked Scylla as poetic creations.

Contemporary with Albertus, the Old Norwegian manuscript, *The King's Mirror,* presents human-shaped sea creatures such as those Olaus learned about from Scandinavian seamen. In a chapter on the marvels of the northern seas, the fictitious father tells his son that, "It is reported that the waters about Greenland are infested with monsters." Creatures of either sex have human facial features. The "merman," slender beneath broad shoulders, is shaped like an icicle. The other "prodigy," called "mermaid," has the form of a woman, with nipples on its breast, much hair, and long fingers. Below the waist is a scaly fish tail. Whenever either monster appeared, sailors were sure a storm would follow. If the beasts turned toward their ship, seamen feared for their lives, but if the creatures dived away, the crew knew they were spared, even though high seas were sure to come. The author does not identify the monsters with the seals and walruses described later in the chapter.

Conrad Gesner's horned sea devil with humanlike torso and body of a fish. It was based on an artist's drawing of the mummified monster that was displayed in Antwerp.

Map Legacy

No variations of Olaus's Sea Creatures appear on Sebastian Münster's *Monstra Marina & Terrestria* or Abraham Ortelius's *Islandia*. The decorative mermaids sporting in the ocean on Ortelius's map of the Indies are clearly conventional figures of the imagination, not of natural history. Conrad Gesner reproduces both of Olaus's Sea Creatures along with the Sea Unicorn, Sea Rhinoceros, and Sea Cow on the concluding page of the *Historiae Animalium*'s section of animals based on the *Carta Marina*.

In his 1635 *Speculum Mundi*, John Swan takes issue with what Olaus wrote of sea creatures "in his 21. Book." Swan does not accept that tempests rise supernaturally when the monsters are caught. He argues rationalistically that those storms occur "namely by the thickening and breaking of the air; which the snortling, rushing, and howling of these beasts, assembled in an innumerable company causeth."

"The Little Mermaid" statue in Copenhagen Harbor, inspired by Hans Christian Andersen's 1837 fairy tale, which generated the best-known of all modern mermaid images. It is a sweet, tragic figure, outside the age-old tradition of treacherous Sirens.

And Since

Captain Richard Whitbourne's 1602 encounter with a strange mermaidlike creature is one of many first-hand reports that are similar to what Olaus had read and heard about humanlike sea monsters. As he tells it in his *Discourse and Discovery of New-Found-Land* (1620), a blue-haired sea beast that swam toward him in St. John's Harbor surfaced again beside his crew's boat. It "did put both his hands upon the side of the boate, and did strive to come in." One of the seamen "strooke it a full blow on the head, whereat it fell off from them." It then approached two more vessels in the port, sending the sailors fleeing to shore. Whitbourne suspects that the strange creature was a mermaid, but he adds: "whether it were a Mermaide or no, I know not; I leave it for others to judge, etc."

Pliny, the anonymous author of *The King's Mirror*, medieval seamen, and Olaus would surely have regarded Whitbourne's story as one more testimonial to the dangers that sea creatures with human forms pose to mariners. In his 1890s *Curious Creatures in Zoology*, John Ashton nostalgically consigns such acceptance to "ages of faith, but now the materialism of the present age would shatter, if it could, our cherished belief in these Marine eccentricities." Beautiful mermaids, for one, had been reduced to "repulsive looking" manatees

A Rosmarus

Map B Ⓑ

Up in Greenland waters, on the final leg of the voyage, the bearded Sea Creature looks toward what the *Carta Marina* key describes as, "Two colossal sea monsters, one with dreadful teeth, the other with horrible horns and burning gaze—the circumference of its eye is 16 to 20 feet." Fearsome in size and powers, the two beasts face each other as in confrontation. Olaus does not identify them by name.

"There are monstrous fish on the Coasts or Sea of Norway, of unusual Names, though they are reported a kind of Whales, who shew their cruelty at first sight, and make men afraid to see them; and if men look long on them, they will fright and amaze them."

The final two beasts to be encountered on the voyage up the *Carta Marina* are especially difficult to match with actual marine animals. As Olaus says in his commentary, the "fish" have "unnatural names" (although he does not specify any) and are described as "a kind of Whales," simply meaning large sea monsters.

The vignette of the *History* chapter source of Olaus's brief commentary pictures both monsters, but the corresponding large figure does not depict the teeth of the first of the pair as it is identified in the *Carta Marina* key. (A smaller figure does.) The chapter, "Of the horrible Monsters of the Coast of Norway," describes the horned beast in detail without specifically mentioning the one with "dreadful teeth." Olaus has less to say about the toothed animal than about nearly any other sea beast on the map.

Like the bearded Sea Creature, the two "colossal" monsters are placed beneath the east coast of what is labeled "Grutlandie Pars," a part of the map's Greenland. Scholar Edmund Lynam notes that, "Absurd though it looks to modern eyes it is better than any representation of Greenland on earlier maps." A western section of Greenland is charted in the upper left-hand corner of the *Carta Marina*.

The Sea Creature peers across the water at the opposing monsters. Neither creature is named, but the one with the fearsome teeth is not described in the History. *The only others that are cited in the* Carta Marina *key but not in Olaus's book are the "monster looking like a rhinoceros" (E e) and "another grisly monster, name unknown" (D e).*

GRVTLANDIE
PARS
B
B
C
CETE
GALLEA
PEREGRI
NA
OSTRAFIORD
MIDELFIORD
REDH
HOLM
TROMS
EGGE
MALANGER
B
HASEVOG
LOKHELLE
VIKFIE
GILLEFIORD
SANDER
GIGASVND
SKARTASVND
VAN
LANGE
QVEDEFIORD
ROLLEN
LANGANES
TRÖDANES
ROST
DVVANES
ANDANES
DOMVS
BIRLARORVM
SCONGEN
HICESTHORRENDA
F
VNDER
VAL
K

A ROSMARUS

Carta Marina

Like the horned monster blocking its northeasterly progress up the *Carta Marina*, the massive striped beast with sharp, upturned tusks has been drawn from the imagination without a stated model. It is described in the map's brief key only as an enormous sea monster with "dreadful teeth." Many of the *Carta Marina* marine figures are depicted with teeth, but because teeth, or tusks, are the most notable feature of this beast, they invite comparison with a like animal on the *Carta Marina*.

The form of Olaus's colossal monster with teeth approximates a walrus in tusks, bulk, flippers, and tail. A somewhat naturalistic—and named—counterpart of that fantastic beast is the Rosmarus (Morse, Sea Elephant, Sea Horse, walrus), which hangs by its teeth from a cliff (B e). Olaus does not note similarities between the two animals, nor does he relate either one to the Sea Swine and Sea Cow—both of which share walrus characteristics.

While only referring to the first colossal monster in a single phrase in the map's key, Olaus details the Rosmarus's nature, capture, and products in his *History*:

> *"The Norway Coast, toward the more Northern parts, hath huge great Fish as big as Elephants, which are called Morsi, or Rosmari, may be they are so from their sharp biting; for if they see any man on the Sea-shore, and can catch him, they come suddenly upon him, and rend him with their Teeth, that they will kill him in a trice ... They will raise themselves with their Teeth, as by Ladders to the very tops of Rocks, that they may feed on the Dewie Grasse, or fresh Water, and role themselves in it, and then go to the Sea again, unless in the mean while they fall very fast asleep, and rest upon the Rocks, for then Fishermen make all the haste they can, and begin at the Tail, and part the Skin from the Fat; and into this that is parted, they put most strong Cords, and fasten them on the rugged Rocks, or Trees, that are near; then they throw stones at his head, out of a Sling, to raise him, and they compel him to descend, spoiled ..."*

Olaus's History *vignette of a Rosmarus that sleeps upon a* Carta Marina *cliff (B e). As* History *authority John Granlund notes, the vignette's depiction of hunting from the sea differs from the text description of the fishermen attaching the cords to rocks or trees to secure the animal.*

A rosmarus (Rostunger, Rosmar) on Abraham Ortelius's Islandia *(N). Although the animal's "teeth" grow from the top of its head, they are pointed downward, unlike those of Olaus's Rosmarus and Conrad Gesner's copy of that figure.*

The Rosmarus is skinned alive as fishermen pull it down to earth, where it dies from loss of blood. Rosmarus teeth are prized for their "hardness, whiteness, and ponderousnesse" and are carved into weapon handles and chess pieces. Their hide makes strong bell ropes and cords for pulleys.

Ancestral Lore

Albeit the Arctic walrus is outside of Mediterranean classical traditions, it naturally figures in accounts of Viking trading of blubber, fur, hides, and ivory.

Olaus's description of the Rosmarus is heavily indebted to that of one of his medieval sources, Albertus Magnus. The ecclesiastical naturalist describes the same method of walrus hunting in detail. He notes that he had seen thongs made of walrus hide in the marketplace at Cologne. In keeping with his time, he includes the walrus in his extensive entry on whales.

According to the anonymous author of the contemporaneous *King's Mirror,* Greenlanders, too, regarded the walrus as a whale. The author would prefer to group the animal (accurately) among the seals, which he calls "fish," because they eat other fish. Walrus tusks grow to an ell and a half (about 5½ feet), and walrus hide makes such strong ropes that sixty men cannot break it.

Map Legacy

Sebastian Münster does not represent a walrus among his sea monsters. Abraham Ortelius does (*Islandia*, N). Similar to the author of *The King's Mirror,* he considers his naturalistic "Rostunger," or "Rosmar," as "somewhat of a sea calf" (seal). Like the walrus of Albertus and Olaus, this creature sleeps while hanging on a rock by its teeth.

While he is skeptical about Olaus's sea monster figures, Conrad Gesner replicates the *Carta Marina* colossal beast as an elegantly striped "boar whale" with upward-curving tusks. Ambroise Paré accompanies his adaptation of the woodcut with text that emphasizes the animal's long "sharpe and cutting" tusks and "scailes set in a wonderfull order."

Conrad Gesner's "boar whale." Gesner attributes the figure to Olaus, as he does other sea monsters from the Carta Marina, *placing responsibility for their accuracy on Olaus himself.*

Gesner elsewhere reproduces the figure of a Rosmarus/walrus from Strasbourg and contrasts it with Olaus's Rosmarus. Renaissance natural history scholar Brian W. Ogilvie (*The Science of Describing*, 2006) points out that while Gesner criticizes the Strasbourg illustration for its rendering of feet on the "fish," he nonetheless regards it, with its downward-curving tusks, as more accurate than Olaus's Rosmarus, whose tusks, pointing upward, would not hang from rocks.

In his *Whale Book,* the amateur naturalist Adriaen Coenen renders Conrad Gesner's Olaus-inspired "boar whale" and Olaus's *History* vignette on different manuscript pages, unaware that they portray the same animal. He attributes both of these strange "fish" to Olaus. Such is Olaus's walrus-like monster with "dreadful teeth" in a transitional age.

The Lewis Chessmen (AD 1150–1200). The history of the medieval pieces carved mostly from walrus ivory is uncertain. Scholars disagree on whether the pieces are of Icelandic, Norwegian, Scottish, or English origin.

And Since

Walrus ivory, too, has a legacy—as chess pieces such as Olaus mentions: the celebrated Lewis Chessmen. After centuries of being unknown, the ninety-three carved pieces were discovered on the Isle of Lewis, Scotland, and made public in 1832. Since then, these rare artifacts have been prized museum exhibits up to our own day.

THE KRAKEN

Map B Ⓑ

Confronting the massive monster with "dreadful teeth" is the final beast to be encountered on the voyage up the *Carta Marina*: another colossal monster, which the map key describes as having "horrible horns and a burning gaze." Its ensuing fame as one of Olaus's most innovative monsters is second only to the Sea Serpent. Olaus continues the commentary he began with the unnamed facing pair.

"Their Forms are horrible, their Heads square, all set with prickles, and they have sharp and long Horns round about, like a Tree rooted up by the Roots: They are ten or twelve Cubits long, very black, and with huge eyes: the compass whereof is above eight or ten cubits: the Apple of the Eye is of one Cubit, and is red and fiery coloured, which in the dark night appears to Fisher-men afar off under Waters, as a burning fire, having hairs like Goose-Feathers, thick and long, like a Beard hanging down; the rest of the body, for the great of the head, which square, is very small, not being above 14 or 15 Cubits long; one of these Sea-Monsters will drown easily many great ships provided with many strong Marriners. The long and famous Epistle of Ericus Falchendorf, Arch-Bishop of the Church of Nidrosus ... which is the Metropolis of the whole Kingdom of Norway, and it was sent to Leo the Tenth about the Year of Grace, 1520, and this confirms this strange Novelty: and, to this Epistle, was faned [enshrined] *the head of another Monster seasoned with Salt."*

Like Olaus and Johannes Magnus, Erik Walkendorf, Archbishop of Nidaros (later Trondheim), Norway, sought exile in Rome during the Protestant Reformation. He became acquainted with Johannes there, two years after sending Pope Leo X the letter describing the wonders of his Finnmark diocese, including sea monsters. The preserved animal head accompanying the letter was that of a walrus, such as that in Albrecht Dürer's 1522 drawing of a walrus head.

The threatening, colossal beast with blazing eyes and horns like an uprooted tree is the final monster obstacle on the voyage up the Carta Marina.

ANDIE
RS
STAPPEN
B
B
HELGA
KEDELVI
SEGERVER
TROMS
REDH
HOLM
MIDELFIORD
EGGE
MALANGER
HASEVOG
LOKHELLE
B
VIKFIE
FELBY
AVVER
GILLEFIORD
OSTRAFIORD
GALLEA
PEREGRI
NA
V
SANDER
GIGASVND
SKARTASVND
VANDVAD
LANGE
QVEDEFIORD
ROLLEN
LANGANES
TRODANES
DVVANES
ANDANES
ROST
DOMVS
BIRCADORVM
BERKARA
QVENAR
VNDER
K
CAPITANE

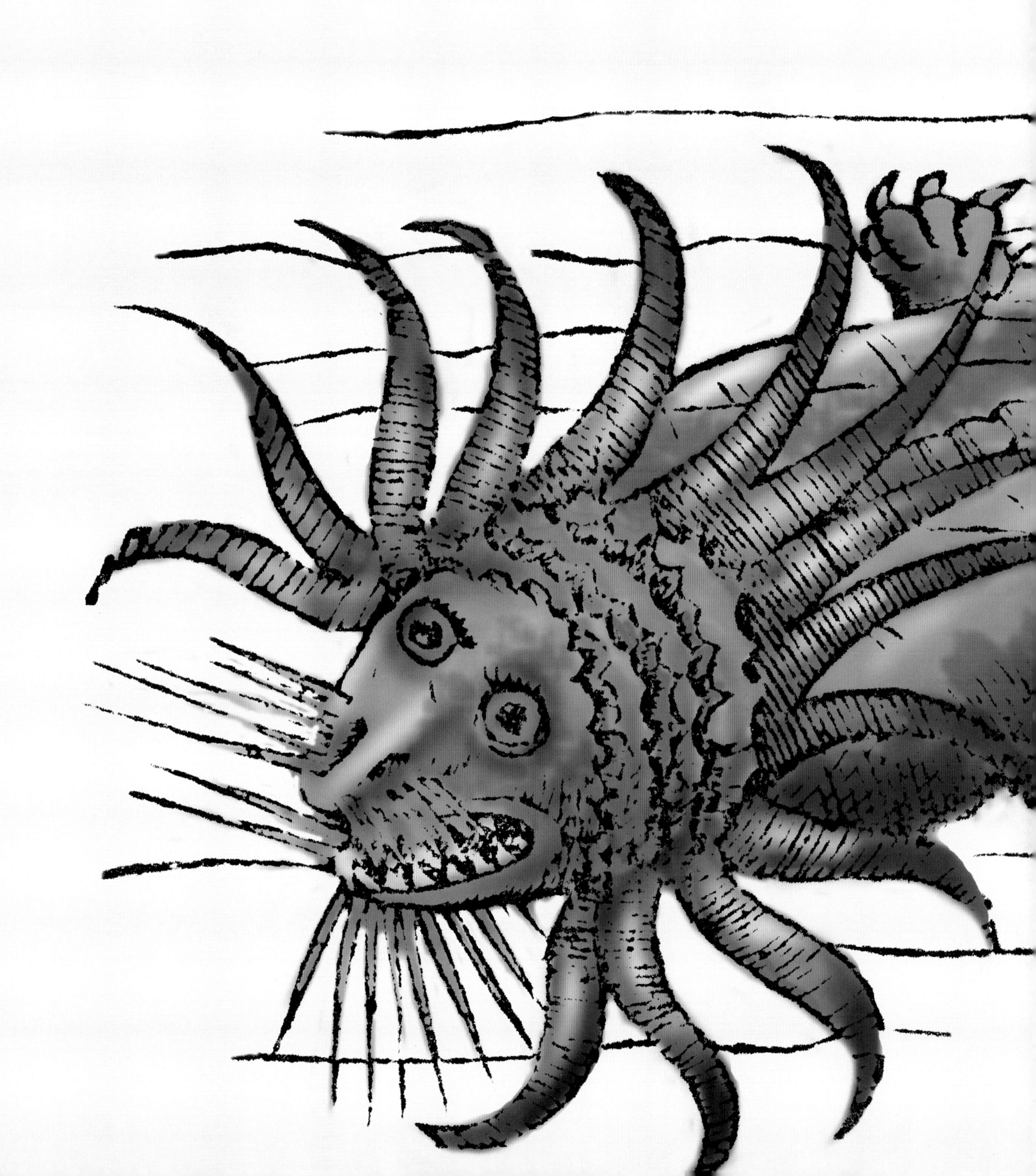

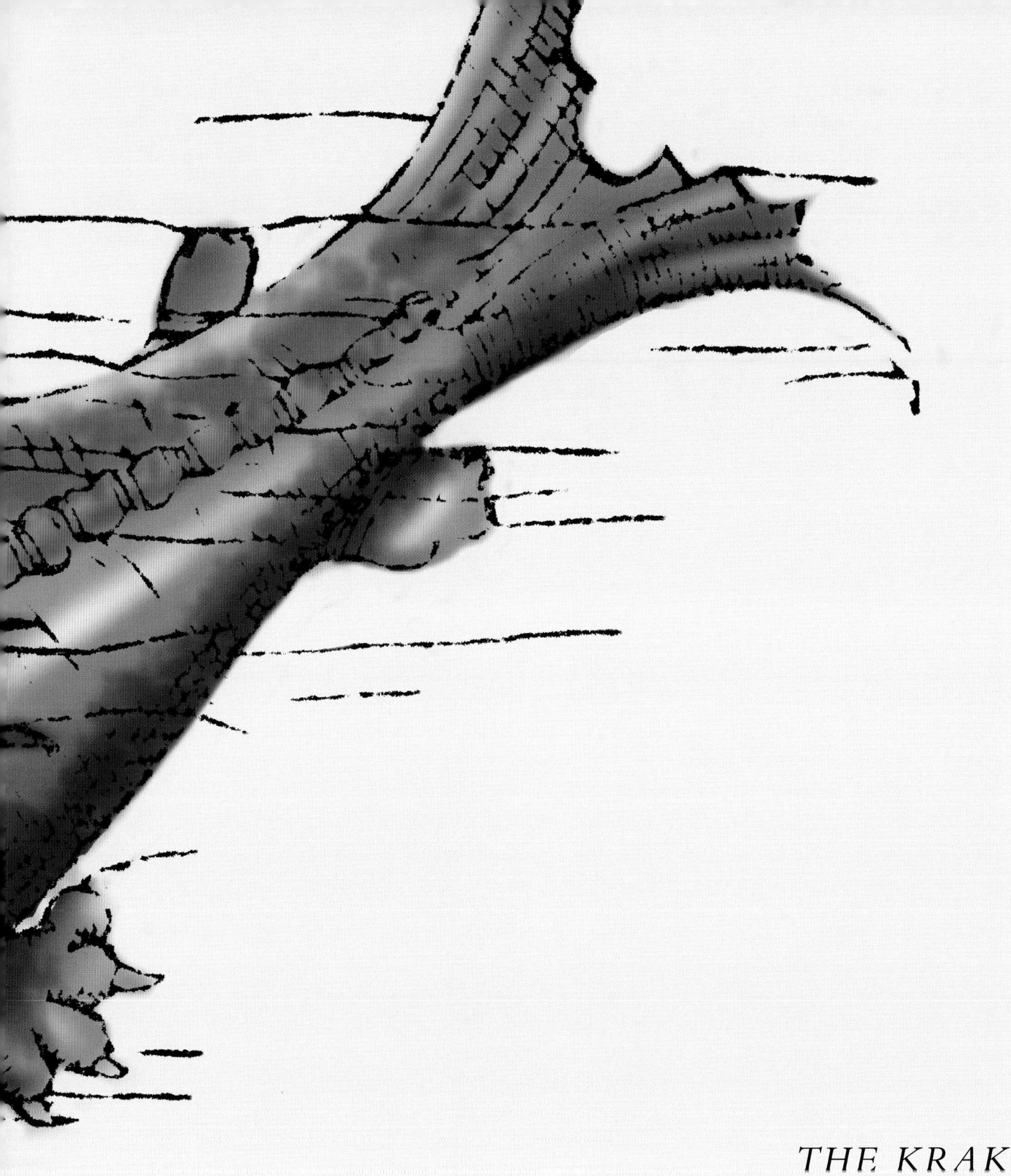

THE KRAKEN

Carta Marina

Adriaen Coenen's horned fish based on Olaus's History *vignette of two colossal sea monsters. Coenen regards the beasts as wonderful but fantastic.*

Olaus Magnus scholar John Granlund calls the *Carta Marina* figures of the two colossal sea monsters "fantasies." And so they are, like other marine beasts on the map. The colossal monster with "dreadful teeth" bears a similarity to the rather naturalistic Rosmarus, hanging from a cliff by its teeth. The "horrible horns" of the monster facing it vaguely relate to the many appendages of the map's Polypus/lobster, which Olaus described as an octopus. If Olaus had an actual marine counterpart in mind for the horned beast—as he did for the rhinoceros-like monster—he does not say so. Although its "terrible horns" and "burning gaze" hint at tentacles and eyes of large cephalopods, it is a particularly enigmatic figure.

It is thus ironic that Olaus's oft-quoted *History* commentary foreshadows the beast's eventual acceptance as one of Olaus's two most celebrated monsters. Its history is long and complex, appropriately entangled with the mythical kraken and associated with the giant squid (*Architeuthis*, "ancient squid"), an animal regarded as fabulous prior to its discovery in the nineteenth century. In chapters on the kraken, sea monster authorities Willy Ley (*Exotic Zoology*, 1941) and Richard Ellis (*Monsters of the Sea*, 1994) accord Olaus an honored early place in kraken/giant squid lore as the first scholar since classical times to describe "the giant cuttlefish."

An octopus on a Minoan stirrup vase (ca. 1500 BC) from the restored Palace of Minos, at Knossos, Crete. This is a replica of the original vase in the Herakleion Museum.

Ancestral Lore

A distant mythical relative of Olaus's horned beast is the second of the pair of monsters evoked by the *Carta Marina*'s Caribdis Maelström: the horrible Scylla.

The monster that Homer's Odysseus must pass to avoid being sucked into the vortex of Charybdis has "twelve misshapen feet, and six necks of the most prodigious length; and at the end of each neck she has a frightful head with three rows of teeth in each." When any ship passed beneath her den, all the necks would dart out like tentacles and she would seize men in her mouths. Scylla thrived in myths of the Mediterranean, an area where cephalopods were a familiar marine animal, savored as a delicacy and a favorite subject of Minoan pottery.

What is thought to be the seminal classical account of a giant cephalopod is, not surprisingly, found in Pliny's *Natural History.* According to the Roman naturalist's authority, Trebius Niger, a large polyp continually raided coastal fishponds at Carteia, Spain. The beast eluded capture until hounds cornered it one night and the keepers harpooned it. Its tentacles, covered with suckers, were 30 feet long, its carcass weighed 700 pounds, and its cask-size head yielded 90 gallons. Pliny himself concedes that Niger's report smacked of the miraculous. Olaus does not cite this incident in his horned monster description, but in his *History* commentary on the Polypus, he did refer to Niger's description of polyps' deadly attacks on sailors and swimmers.

The medieval north had its own sea monster, the *hafgufa*, which evolves into what is called the mythical kraken. It, in turn, becomes associated with the giant squid. In an Old Icelandic romance, *Örvar-Odds* ("Arrow-Odd"), the *hafgufa*, the largest of all sea monsters, swallows ships, men, and whales. A second monster in the saga is the *lyngbakr,* an island whale that sinks with sailors on its back.

Sebastian Münster's friendlier version of Olaus's monster with terrible horns.

The Old Norwegian *King's Mirror* (ca. 1250)—the treatise that classifies the marine life of the Icelandic seas—also describes the *hafgufa*. The anonymous author is cautious, even skeptical, in his serious consideration of a fish "which is scarcely advisable to speak about on account of its size." The rare times it has been seen, "it has appeared more like an island than a fish." The only thing told about its habits is that its belching of food draws fish into its open mouth. "In our language," the author notes, "it is usually called the 'kraken.' "

This hearsay monster, conflated with the *lyngbakr*, is clearly derived from the Aspidoceleon of the *Physiologus*, in spite of its not having sweet breath. The historian Laurence Marcellus Larson's 1917 rendering of *hafgufa* as "kraken" presumably follows an oral tradition before Erich Pontoppidan establishes the word in print in the eighteenth century. Larson's "kraken" usage represents the fluidity and complexity of the kraken, which lived in some form in the vast body of lore that Scandinavian fishermen shared with Olaus.

Map Legacy

On *Monstra Marina & Terrestria*, Sebastian Münster transforms Olaus's unnamed monster with "horrible horns" into a charming, nonthreatening figure swimming away from the beast with "dreadful teeth." The key (D), however, repeats Olaus's key, with the additional details of a square head, "a great, long beard," and a body smaller than the head. Olaus's fantastic beast is not pictured on Abraham Ortelius's *Islandia*. Adriaen Coenen varies Olaus's *History* vignette with the puzzled comment that, "You can't represent these fish as strangely as Olaus describes them. Olaus writes very wonderful

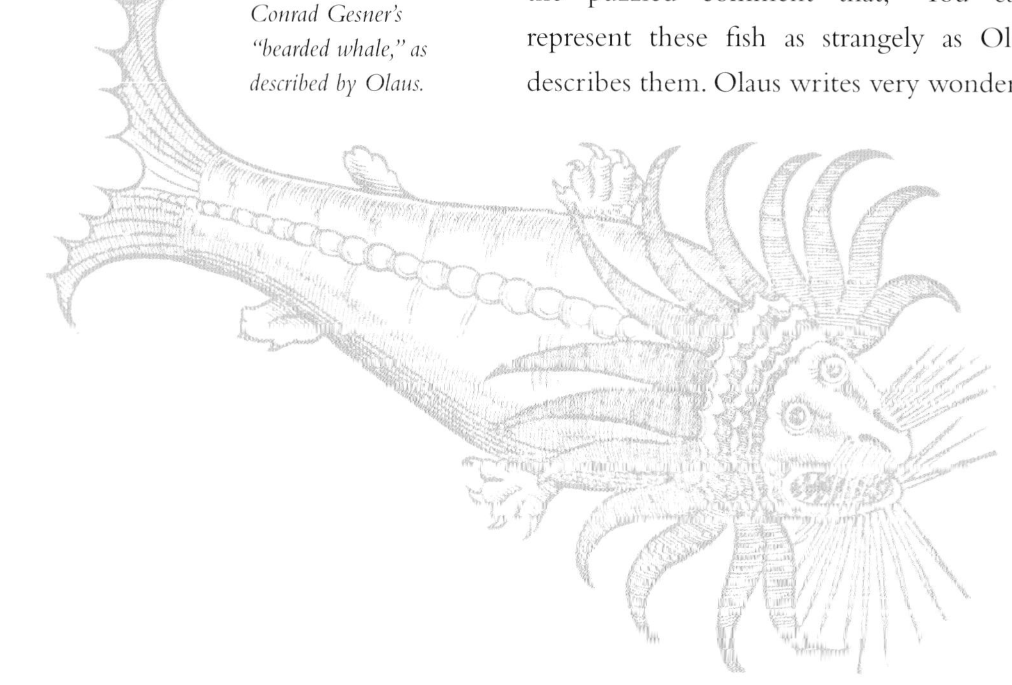

Conrad Gesner's "bearded whale," as described by Olaus.

things about these fish, but they are not credible." Only Conrad Gesner ventures to identify the *Carta Marina* monster, calling it a "bearded whale."

And Since

Exactly 200 years after the publication of Olaus's *History*, Erich Pontoppidan, Bishop of Bergen, names a monster similar to Olaus's horned beast: "Kraken, Kraxen, or, some name it, Krabbern." He does not mention the *Carta Marina* and *History* monster in *The Natural History of Norway* (1755), but as an author of the Enlightenment, he dismisses the island whale story of the "credulous Ol Magnus" as a "notoriously fabulous and ridiculous romance." He generously regards as a printer's error Olaus's description of "the Norwegian Sea-snake" (Sea Worm) as being 80 feet long but only as thick as a child's arm. And he excuses Olaus for mixing "truth and fable" in his account of the Sea Serpent due to his living in a dark age of superstition. Pontoppidan does, however, concede that, "we in the present more enlighten'd age are much obliged to him, for his industry, and judicious observations."

Bishop Pontoppidan introduces his chapter on "certain Sea-monsters, or strange and uncommon Sea-animals" with rationalistic caution. After discoursing on European mermen and the sea serpent, he arrives at "the third and incontestably the largest Sea-monster in the world": the kraken. The back of Pontoppidan's monster appears to be "about an English mile and an half in circumference." When the great beast surfaces, its expanse resembles a group of small islands. Then its "horns" appear, arms or tentacles as tall as masts of midsize ships. Its "peculiar" scent attracts fish into its mouth. Its sinking into the sea spins the waters into a whirlpool that pulls down everything near it. Pontoppidan even contends that the remora, known for stopping a ship in mid-sail, is actually a kraken.

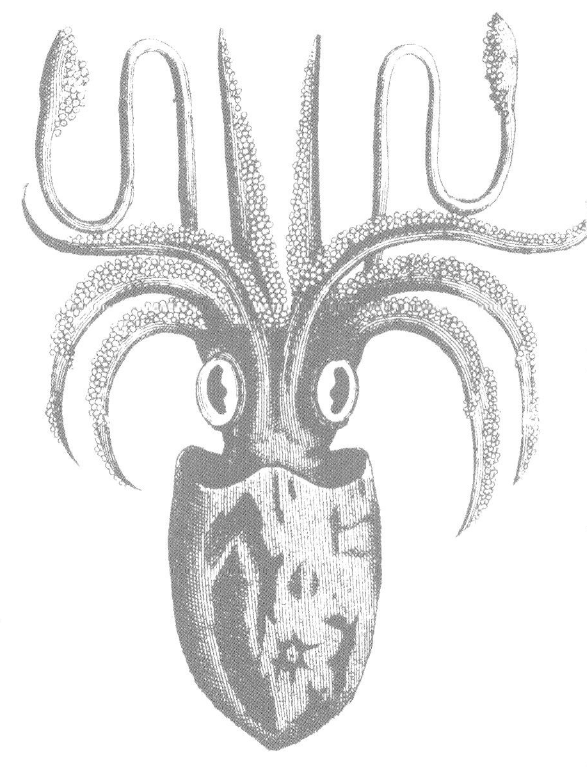

A squid from Matthäus Merian the Younger's Theatrum universale omnium Animalium *(ca. 1650). The large eyes and hornlike tentacles of this natural history specimen bear a striking resemblance to Olaus's colossal monster.*

An artist's vision of the entire giant squid that the crew of the French warship Alecton *attempted to capture near Tenerife in 1861, from* L'Univers Illustre *(ca. 1867).*

Pontoppidan's kraken embodies similarities with more *Carta Marina* figures than just the horned monster of the map and the *History*. Like the "kraken" of *The King's Mirror,* this beast's island appearance and allurement of small fish with its breath evokes the *Physiologus* source of Olaus's Island Whale. This kraken's creation of a great vortex resembles the *Carta Marina*'s Caribdis/Maelström, associating Olaus's horned monster with both Scylla and Charybdis. And the kraken's arms, as tall as masts, are reminiscent of the Sea Serpent. Sea monster experts Henry Lee (*Sea Monsters Unmasked*, 1883) and Richard Ellis both suggest that the sea serpents of many documented sightings were actually tentacles of giant cephalopods.

Half a century following the publication of Erich Pontoppidan's natural history, the French naturalist Pierre Denys de Montfort publishes a zoological study in which he notes the Norse *hafgufa* and "*lyngback*" and discourses at length on the kraken of Pontoppidan and the sea beasts of Olaus Magnus. However, his exaggerated accounts of colossal kraken and octopuses discredit his 1802 *Histoire naturelle générale et particulière des mollusques* and contribute to the loss of public belief in such fantastic stories. By 1822, Sir Walter Scott makes references to "the wondrous tales told by Pontoppidan."

Disbelief in colossal cephalopods was challenged in 1861 when a French warship, the *Alecton*, encountered a giant squid near Tenerife, in the Canary Islands. The captain ordered the crew to fire upon the 24-foot animal in order to capture it. With bullets and harpoons failing to harm it, the crew looped a rope around its tail and attempted to haul it aboard. The body of the squid broke off, leaving the tail as evidence of the partial capture—and of the existence of monsters such as those described by Olaus, Pontoppidan, and de Montfort.

Jules Verne uses both de Montfort's book and the *Alecton* incident in turning the actual into science fiction in his 1870 *Twenty Thousand Leagues Under the Sea*. Preceding the most famous fictional and cinematic battle with a giant squid, Professor Arronax and the other prisoners inside Captain Nemo's *Nautilus* notice a commotion in seaweed outside the large submarine window. The professor explains that they are looking at caverns of devilfish and proceeds to sum up the fantastic history of the actual sea beasts:

> *. . . when it is a question of monsters, the imagination is apt to run wild. Not only is it supposed that these poulps can draw down vessels, but a certain Olaus Magnus speaks of an octopus a mile long that is more like an island than an animal. It is also said that the Bishop of Nidros was building an altar on an immense rock. Mass finished, the rock began to walk, and returned to the sea. The rock was a poulp. Another Bishop, Pontoppidan, speaks also of a poulp on which a regiment of cavalry could maneuver.*

The most famous of all kraken/giant squid illustrations, from Pierre Denys de Montfort's 1802 Histoire naturelle générale et particulière des mollusques. *The hand-colored woodcut is a reproduction of art in the Church of St. Malo in France.*

Illustration from Jules Verne's 1870 Twenty Thousand Leagues Under the Sea. *Professor Arronax gazes in wonder at a giant squid outside Captain Nemo's* Nautilus.

Along with Pontoppidan's "poulp," Olaus's horned monster has a name here: octopus, although, in the manner of folklore, the beast is conflated with his Island Whale. The Bishop of Nidros is Erik Walkendorf, whose letter Olaus mentions in his seminal commentary on the kraken/giant squid.

~

The *Carta Marina* altercation between the two colossal monsters remains unresolved as the voyage continues northeasterly, past a Rosmarus sleeping on a cliff, to safe anchorage in a Finmarchia harbor, on the edge of the Vast Ocean.

LANDFALL

The sea monster images that attract the eye of someone looking at the *Carta Marina* for the first time dominate this book, bursting out on double pages. Thanks in great part to Conrad Gesner's woodcuts of Olaus's figures, images of the beasts have been scattered throughout popular books since the nineteenth-century revival of interest in fabulous animals. And the iconic woodcuts are now ubiquitous on the Internet. The pictures are even more central to Olaus's enduring sea-monster legacy than his oft-quoted *History* commentaries on the Sea Serpent and what others have identified as the kraken or giant squid.

It was, though, the influence of the *Carta Marina* beasts on the two other preeminent sea monster maps, with identifying keys, that eventually resulted in the voyage of this book and docking in Olaus's Finmarchia.

While I was researching other imaginary animals, the first of the two maps whose strange beasts fascinated me was Abraham Ortelius's *Islandia*. It was a frontispiece, with appendix key, in Muriel St Clare Byrne's *The Elizabethan Zoo: A Book of Beasts Fabulous & Authentic* (1926). Her book is a loose collection of chapters from Edward Topsell's natural histories and Philemon Holland's 1601 translation of Pliny's *Natural History*. The accompanying woodcuts are either Topsell's or Conrad Gesner's.

Then there was the dust jacket of Willy Ley's 1968 *Dawn of Zoology*, which traces the evolution of natural history from Aristotle to Charles Darwin. The sea beasts on that book cover were more fanciful—and sensational—than the figures on the Ortelius map. Spouting monsters threatened a ship, and a sea serpent coiled around another vessel. The figures on this plate were also keyed, as on *Islandia*, but neither the book nor the jacket attributed the source of the chart.

"Very like a whale," Shakespeare's Polonius said, agreeing with Hamlet's description of a cloud. In Olaus's time, "whale" meant any large sea monster. His influential depiction of cetaceans evolves over the centuries. Above: An engraving of a pod of spouting whales. Below: Olaus's Pristers are distracted with trumpet blasts and bobbing casks.

I later discovered the chart was Sebastian Münster's *Monstra Marina & Terrestria*. Willy Ley, a "romantic naturalist," oversaw the publication of facsimile volumes of Topsell's first English translation of Gesner's natural history.

A children's book reproduction of Antonio Lafreri's 1572 version of Olaus's *Carta Marina* led me to the 1539 map, its influence on the Ortelius and Münster maps, and to Olaus's *History of the Northern Peoples*.

Olaus's total map of Scandinavia illustrates the West's grand transition from medieval thought to modern inquiry. As the largest and most accurate map of Scandinavia of its time, the *Carta Marina* contributed greatly to Renaissance cartography. At the same time, its hundreds of figures on the terrestrial portions of the map grew out of the medieval Mappa Mundi tradition. Most of the monsters in the northern seas of the *Carta Marina* are the imagination's shaping of age-old oral and written traditions—even as they influence nascent zoology.

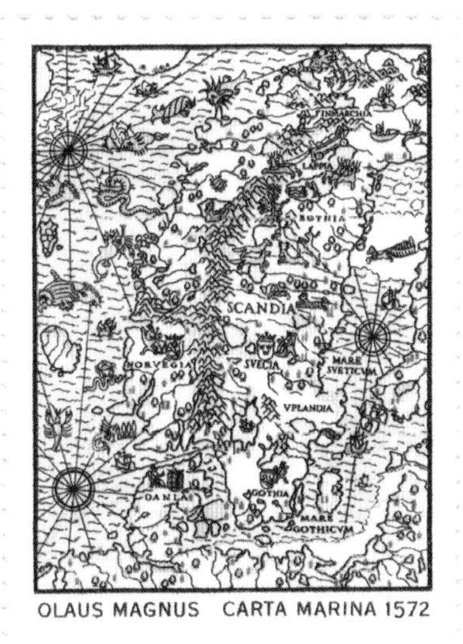

Swedish stamp of Antonio Lafreri's 1572 version of Olaus Magnus's Carta Marina. *A 1991 affirmation of Olaus's enduring legacy, four centuries after his map was printed.*

That the fantastic sea beasts are meant to correspond to actual animals is evident in their keys, which are presented as matter-of-factly as those that identify regions, history, and customs of the peoples of Scandinavia. Olaus affords the monsters even more credibility by the vignettes and chapters he devotes to them in the climactic Book 21 of his *History of the Northern Peoples*.

Naturally enough, Olaus's *History* is like the *Carta Marina* in straddling two intellectual worlds. On the one hand, Olaus is a typical early Renaissance scholar in his dependence on Pliny and other classical and medieval authorities. On the other hand, the last Archbishop of Scandinavia infuses his writings with information and lore gathered from his travels throughout much of the North as a churchman. His commentaries on individual animals are a blend of scholarship, oral tradition, and personal experience—and are sometimes misguided, contradictory, and confusing. They are a reflection of his transitional age.

Olaus's sea monsters clearly indicate the fluid nature of natural history. Conrad Gesner, widely regarded as the Father of Modern Zoology, included in his immense *Historiae Animalium* all that was known of any given animal, whether it was myth, folklore, or observed reports from trusted contemporaries. He was thus a perfect foil for Olaus. He reproduced all but one (the Rockas) of Olaus's *Carta Marina* beasts—with the disclaimer that Olaus alone was responsible for them and the information. Gesner's woodcuts of Olaus's monsters were widely disseminated through subsequent natural histories. One of those was the manuscript of amateur naturalist Adriaen Coenen, whose charming *Whale Book* was a happy discovery of my research.

The influence of Olaus's *Carta Marina* monsters on cartographical and natural history giants of the age, namely Sebastian Münster, Abraham Ortelius, and Conrad Gesner, certainly earned him the approbation of marine-life authority Richard Ellis: "Olaus Magnus (1490–1557), the Catholic archbishop of Sweden, was influential as a historian and a cleric, but his name will be forever associated with sea monsters."

CARTA MARINA

Full Key

This is adapted from the University of Uppsala Library's complete English translation of the Latin key in the lower left-hand corner of Olaus's *Carta Marina*:

Carta Marina et descriptio septemtrionalium terrarum ac mirabilium rerum in eis contentarum diligentissime eleborata anno dni 1539

~

A marine map and description of the northern countries and their remarkable features, meticulously made in the year 1539.

"Olaus Magnus Gothus greets the honorable reader. To be able to understand easily this map of the Scandinavian countries and the wonders that exist there (I published this edition in honor of the revered Doge Pietro Lando and the senate of Venice and for the general benefit of the Christian world), you must know that it is divided into nine parts, indicated by the large letters A, B, C, etc, and further that there are several smaller letters within the area of each big letter, which indicate matters briefly referred to in the commentary."

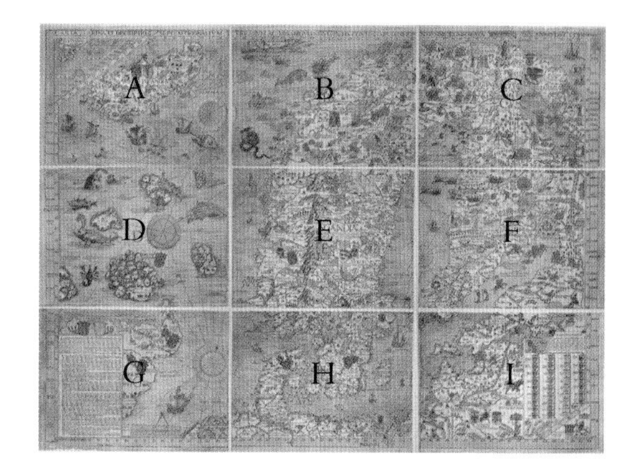

The Carta Marina *is divided into a nine-part grid marked from A to I. A later colored version of the map is printed inside the book's fold-out jacket. The original black-and-white 1539 map is shown on pages 8–9.*

A—Iceland; part of Greenland

A comprises the island of Iceland (Islandia), renowned for its unusual wonders; on this island, close to the small letter (*A*) ~ there are three mountains, their highest peaks glistening with eternal snow and their bases flaming with eternal fire. (*B*) ~ Four springs of very different nature: the first one by means of its eternal heat changes everything thrown into it into stone, while preserving the original shape, the second one is intolerably cold, the third one produces "beer," the fourth one breathes forth destructive contagion. (*C*) ~ A fire consuming the water but not burning the wick. (*D*) ~ White ravens, falcons, magpies, bears, wolves, and hares; yet there are also totally black wolves. (*E*) ~ The ice sounding like howling human voices and clearly indicating that human souls are being tormented there. (*F*) ~ A piece of rock that seems to fly through a great cloud of vapor. (*G*) ~ A vast number of fish heaped up for sale in a pile as high as the houses under the open sky. (*H*) ~ An incredible amount of butter. (*I*) ~ The pasture is so lush that unless the cattle are kept from grazing, they are destroyed through overfeeding. (*K*) ~ Sea monsters, huge as mountains, capsize the ships if they are not frightened away by the sounds of trumpets or by throwing empty barrels into the sea. (*L*) ~ Seamen who anchor on the backs of the monsters in belief that they are islands often expose themselves to mortal danger. (*M*) ~ Merchant ships attacking each other with cannons in their fight to be first in harbor and do good trade. (*N*) ~ The arms of Norway (Norvegia) and Iceland (Islandia).

B—Another part of Greenland; Finland; Lappland

B comprises first a part of Greenland (Grutlandia), whose inhabitants, shown near letter (*A*) ~ are pictured as very skilful sailors, using leather boats which are safe no matter what danger might threaten; using these they attack and sink the ships of the foreigners. (*B*) ~ Two colossal sea monsters, one with dreadful teeth, the other with horrible horns and burning gaze ~ the circumference of its eye is 16 to 20 feet. (*C*) ~ A whale rising up and sinking a big ship. (*D*) ~ A worm 200 feet long wrapping itself around a big ship and destroying it. (*E*) ~ Rosmarus, a sea elephant, sleeps hanging from the cliff and is caught thus. (*F*) ~ Several horrendous whirlpools in the sea. (*G*) ~ The insatiable and gluttonous wolverine emptying its stomach by squeezing itself between trees. (*H*) ~ A fisherman beating on the ice with a club so as to stun and catch the fish under it. (*I*) ~ Reindeers are tamed flockwise and surpass the swiftest horses when they are put before wagons. (*K*) ~ Demons, who have assumed bodily shape, are serving the people. (*L*) ~ A flock of domesticated reindeers giving milk for household use. ★ A find of gold.

C—The North Pole; regions of Scricfinia and Biarmia

C shows, at letter (*A*) ~ forest people who attack the sailors by night but are nowhere to be found during the day. (*B*) ~ Pagan worshippers adore as a divinity a piece of red cloth attached to the top of a pole. (*C*) ~ Starchaterus, a Swedish fist-fighter, very famous in ancient time throughout all Europe. (*D*) ~ A magnetic island, 30 miles from the Pole, beyond which the sailor's guide, the so-called compass, loses its power. (*E*) ~ A huge eagle wraps its eggs in the flayed skin of a hare; by means of its life-giving warmth the chicks hatched. (*F*) ~ The great white lake (the White Sea), where are to be found fishes and birds of an uncountable number of species. (*G*) ~ A marriage ceremony among the pagan worshippers performed with fire struck from flints held over the heads of the bridal pair. (*H*) ~ Exchange of necessities without the use of money. (*I*) ~ Battle between two kings, one of whom fights with reindeers and foot-soldiers on curved lengths of wood (that is, skis) and carrying bows. He defeats the other who fights on horseback. (*K*) ~ Reindeers draw a wagon (sledge) across the snow and ice. (L) ~ Sealhunting on ice-floes and an incredible abundance of salmon and pike. (M) ~ Marten, sable, ermine, different kinds of squirrels, everywhere an enormous number of beavers. (N) ~ Moscovite merchants dragging their boats between the lakes to do barter trade.

D—The Western Islands*

D shows near the small (A) the island and the diocese of the Faroe Island (Faren), its fish-eating inhabitants cut up and divide among themselves the big sea animals thrown up by the storms. (B) ~ Here heads of ravens are given as a tribute to the governor of the region as a sign that they have killed the destructive bird which kills sheep and lambs. (C) ~ At the approach to this island there is a high rock, which the sailors call the Monk—an excellent protection against storms. (D) ~ The terrible sea-monster Ziphius devouring a seal. (E) ~ Another grisly monster, name unknown, lurking at its side (that is to say at the side of Ziphius). (F) ~ Here lies the island of Tyle (Tile). (G) ~ The Hetlandic (Shetlandic) island and bishopric, a fertile country with the most beautiful women. (H) ~ The Orcadic (Orkney) island and bishopric, 33 in number, which in ancient times was called a kingdom. (I) ~ Ducks being hatched from the fruit of the trees. (K) ~ A sea monster similar to a pig.

**Note: D l through D p are missing from the Latin key but are included in the Italian and German versions. The figures that are omitted but are accompanied by legends on the Carta Marina are: D l—Balena and Orca; D m—a Polypus; D n—the Hebrides; D o—the Prister; and D p—Spermaceti.*

E—Norway and Sweden

E comprises at (A) the name of the island of Scandia, from which in bygone days the most powerful people have gone out into the whole world. (B) ~ The arms of the Kingdom of Sweden (Suecia): three crowns. (C) ~ The arms of the Kingdom of Norway (Norvegia): a lion armed with a broad-axe. (D) ~ Here they try to measure the unfathomable depths of the sea. (E) ~ A monster looking like a rhinoceros devours a lobster that is 12 feet long. (F) ~ Plates are fastened as shields to the feet of the horses so that they will not sink down into the snow. (G) ~ Domesticated reindeers give excellent milk. (H) ~ Lynxes devour wild cats. (I) ~ An attack by wolves against elks on the ice. (K) ~ Pyramids and enormous stones bearing the deeds of ancestors described in Gothic letters. Under the arms of the kingdom of Sweden are the signs for finds of iron, finds of copper, and hoards of excellent silver. (L) ~ A lake that never freezes. (M) ~ A sea snake, 30 or 40 feet long.

F—Finland; Moscow

F Below small (A): it often happens that the sea freezes and is able to carry very heavy carriages (sledges) and at the same time alongside their route leaves a navigable waterway to the seafarers, who compete with the carriages in swiftness. (B) ~ Onagri ("wild donkeys") or elks pull the carriages swiftly over the snow. (C) ~ A fight between shepherds and snakes. (D) ~ Pheasants or woodcocks lie hidden under the snow without food for several months at a stretch. (E) ~ Other birds, totally white snow birds, which never show themselves except during severe winters. (F) ~ A black river of immeasurable depth; it contains only black fishes, but these have a delicious taste. (G) ~ Outbreak of an unendurable din, when something living is thrown into the Viborgian cave or hollow. (H) ~ The den of beavers, partly on land, partly in the water; how they build it by drawing together pieces of wood. (I) ~ The pelican, a bird big as a goose giving out a very strong call with his water-filled throat. (K) ~ The otter is tamed to fish and to bring the fish to the cook. (L) ~ People are moving rapidly over the endless frozen sea with (skates made of) bones under their feet. (M) ~ Vessels for feasting, called Kåsa (Kosa). (N) ~ During the winter fighting takes place on the ice as during summer it does at sea.

G—Scotland, England, the Latin Opera Breve

G gives the whole key to the map as well as parts of the kingdoms of England (Anglia), Scotland (Scotia) and Holland (Hollandia).

H—Denmark; Sweden

H contains at letter (A) the old kingdom of Frisland (Frisia), where there are excellent horses, and then the kingdom of Denmark (Dania), broken up into its many island and among its warlike inhabitants. (B) ~ The mighty and stately Vendic towns, the harbors of which are permanently lit up so that the seamen can avoid running into any danger through carelessness. (C) ~ Public lodgings are sometime built on the frozen sea. (D) ~ Collecting amber on the Prussian coast. (E) ~ The town of Danzig, inhabited by well-to-do and honest citizens. (F) ~ The benevolence of the fishes called rockas in Gothic and raya in Italian: They protect the swimming man and save him from being devoured by the sea monsters. (G) ~ The kingdom of Gothia, the first fatherland of the Goths. (H) ~ The island of Gotland (Gotlandia), according to the etymology of its name, the island of the Goths, where even today cases involving maritime law are settled. (I) ~ Fires are lit on the coastal mountains in wartime. (K) ~ The royal city of Stockholm (Stocholmia), well protected by the art of fortification, by natural formations and water. (L) ~ Mighty ships for sea battle, equipped on all sides with big cannons, from which pieces of iron are fired, enclosed in barrels.

I—Russia; shields of the tribes of Scandinavia

I first contains at the letter (A) ~ the country of Livonia, which is under the rule of the German Order of the Blessed Virgin. (B) ~ Kurland (Terra Curetum), on the coast of which there are repeated shipwrecks and scant comfort is given to the victims. (C) ~ Samogethia, so called after the settlement there of the Goths. (D) ~ The Grand Duchy of Lithuania (Lituania), under the King of Poland. (E) ~ shows an aurochs that easily lifts and tosses a man in full armor. (F) ~ Bears poking honey from the trees, are being beaten down by iron spiked clubs that have been hung there. Finally a table gives the names of a number of peoples who according to the unanimous testimony of the ancient authors originate from the island of Scandia.

GLOSSARY

Sea Monster Counterparts

Modern marine zoology was just beginning to emerge from classical and medieval lore when Olaus Magnus produced his *Carta Marina* and compiled its commentary, *The History of the Northern Peoples*. It is not always possible to identify the real-life prototypes of Olaus's sea creatures due to their pre-Linnean names and imaginative art. In those cases, we can only speculate. Many of them, eons-old wonders of nature rather than of the imagination, are now endangered and protected species.

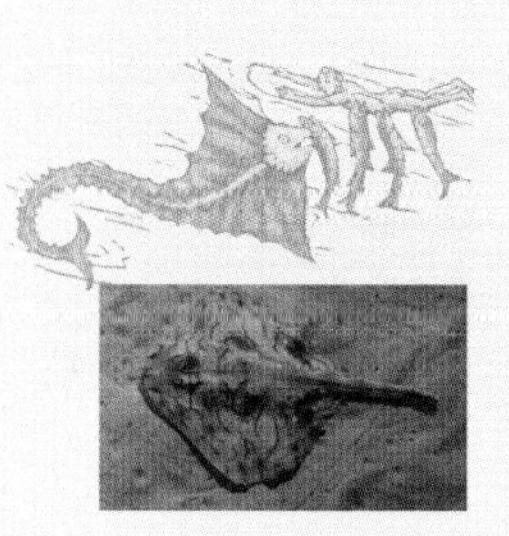

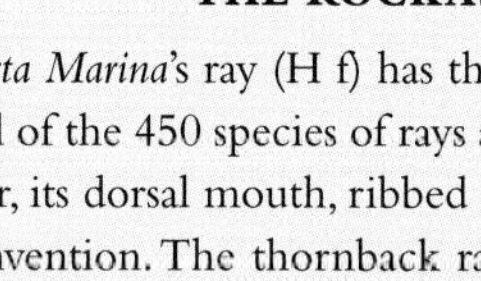

THE ROCKAS / RAY

The *Carta Marina*'s ray (H f) has the winglike pectoral fins and spiny tail of the 450 species of rays and skates around the world. However, its dorsal mouth, ribbed wings, and fluked tail are an artist's invention. The thornback ray *(Raja clavata)* of northern waters grows to 3 feet in length. Olaus's small sharks are probably spiny dogfish *(Squalus acanthias)*.

THE SEA WORM / EEL?

Olaus's "sea snake, 30 or 40 feet long" (E m), is bluish gray, thin, and toxic. Edward Topsell follows Conrad Gesner in calling it a serpent, and Adriaen Coenen identifies it as an eel. One of 730 species of eels, the snakelike European eel *(Anguilla anguilla)* is nearly black in color, weighs up to 29 pounds, and reaches about 3 feet in length.

THE DUCK TREE / BARNACLE GOOSE

The miraculous trees on the Orcades shore of the *Carta Marina* (D i) contribute to the centuries-long legend that ducks or geese are born from trees or driftwood. The actual barnacle goose *(Branta leucopsis)* breeds in Spitzbergen and Greenland, and its black goose relative, the brent goose *(Branta bernicla)* breeds in Arctic regions. Both visit the northern European regions in the winter.

THE POLYPUS / LOBSTER

The lobster between the *Carta Marina*'s western islands (D m) is probably a gigantic version of the 2-foot long European, or common, lobster. The *Homarus gammarus* is one of the few lobster species worldwide that is harvested commercially as a luxury food. Found in the Eastern Atlantic from Norway to the Mediterranean, it is closely related to the American lobster. It is more distantly related to barnacles.

BALENA & ORCA / NORTHERN RIGHT WHALE?

As influential as Olaus's imaginative whale figures were, it is difficult to identify the actual species being depicted. A probable prototype for the *Carta Marina*'s Balena (D l) is the northern right whale *(Eubalaena glacialis)*, which grows to nearly 60 feet in length and weighs up to 90 tons. Right whales are an endangered species, unlike the far-ranging predatory killer whale *(Orcinus orca)*.

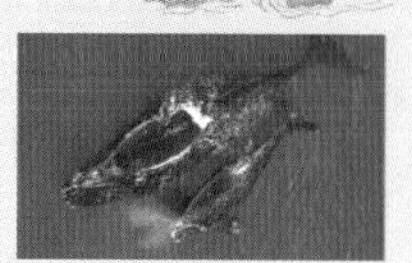

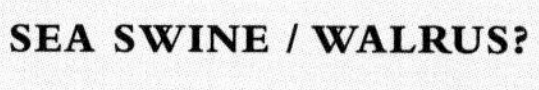

SEA SWINE / WALRUS?

A marine prototype for the fantastic, allegorical Sea Swine (D k) is problematical at best. One could be the tusked walrus *(Odobenus rosmarus)*. Its feeding on the seabed conforms to tradition. Another candidate is the snub-nosed harbor porpoise *(Phocoena phocoena)*. "Porpoise" derives from *porcus marinus,* and the animal is known as the "puffing pig" due to the sound it makes when it surfaces.

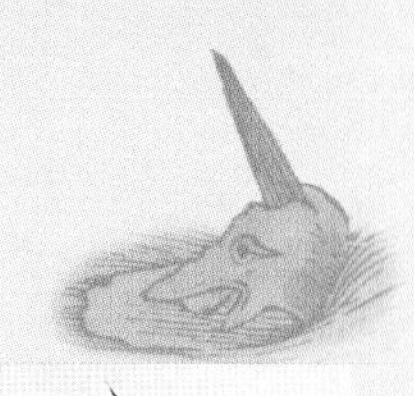

SEA UNICORN / NARWHAL

Olaus had never been to Iceland or Greenland, but his Sea Unicorn (Southwest A, not keyed) represents the narwhal *(Monodon monoceros)*, a cetacean of primarily Arctic waters. Reaching about 16 feet in body length and weighing up to 1¾ tons, the narwhal is best known for its ivory tusk. The remarkable straight and spiraled tooth of the male narwhal grows to nearly 10 feet long.

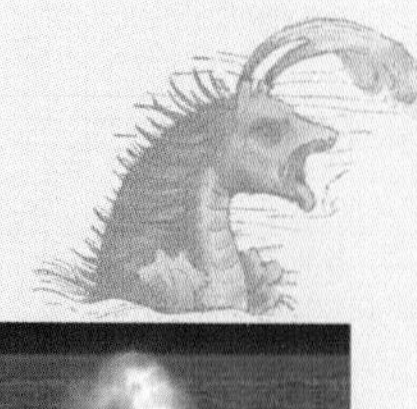

PRISTER / SPERM WHALE

The figure of Olaus's "prister or physeter" (D o) is a sea monster fantasy. The sperm whale (*Physeter macrocephalus* or *Physeter catodon*) is one of its counterparts. Largest of the toothed whales, the male grows to about 60 feet in length and weighs up to 2 tons. Its cloudy spout of exhaled moisture from a single blowhole can reach 16 feet. It is now a protected species.

THE ZIPHIUS / ORCA, KILLER WHALE

The "sword" of Olaus's "swordfish" is a dorsal fin, not the long spearlike nose of the famous game fish *(Xiphias gladius).* Even though the *Carta Marina* Ziphius (D d) is a figure of fantasy, its prominent "sword" and seal victim suggest that its marine prototype is the carnivorous orca/killer whale *(Orcinus orca),* which attacks the Balena and her calf near the *Carta Marina*'s mythical island of Tile (D l).

THE SEA COW / WALRUS?

A literal image of a cow in the sea does not help to identify the marine counterpart of the *Carta Marina*'s Sea Cow (West E, not keyed). While "sea cow" now commonly refers to sirenians, such as the manatee and the dugong, the Dutch word for walrus meant both "sea horse" and "sea cow," two of the sea beasts cited in Olaus's Sea Cow commentary.

THE SEA RHINOCEROS / INDIAN RHINOCEROS?

The *Carta Marina*'s fantastic "monster looking like a rhinoceros" (E e) differs greatly from its terrestrial counterpart. The two are similar, however, in their size, horns, and "speckled" bodies. The Indian Rhinoceros *(Rhinoceros unicornis)* can grow to 12 feet in length and weigh nearly 2½ tons. Its single horn, composed of keratin hairs, is up to 2 feet long. The animal has been hunted nearly to extinction.

A BEACHED WHALE / STRANDED WHALES

The *Carta Marina*'s Faroe Island whale (D a) is one of the earliest printed images of a scene that becomes a popular pictorial subject, following the 1598 beaching of a whale on the coast of Holland. It is estimated that thousands of whales and other cetaceans are stranded on coasts every year—the probable reasons ranging from storms at sea to faulty echolocation. No definitive cause for beaching has yet been determined.

MORE PRISTERS / SPERM WHALE?

The influential *Carta Marina* image of deterring whales with a trumpet and empty barrels (A k) illustrates a later proverb more accurately than it does actual marine situations. While bull whales were known to attack ships in the nineteenth century, such aggressiveness is now rare. Occasional reported "attacks" on yachts are more likely to be the breaching of juveniles. Widespread regulations prohibit the "harassing" of whales.

ISLAND WHALE / BLUE WHALE

The prominent feature that legendary island beasts have in common is size. The largest animal ever to live on the earth is the blue whale *(Balaenoptera musculus).* A baleen whale can consume up to 10 million crustaceans a day. It has been known to grow to 110 feet and to weigh about 200 tons. Hunted to near extinction, it is now internationally protected.

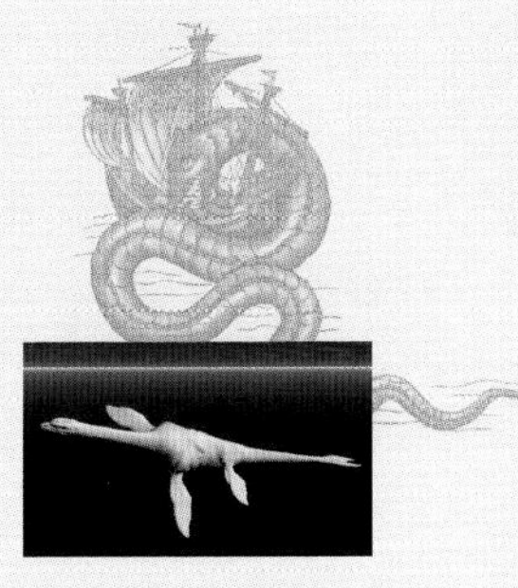

SEA SERPENT / ?

Hundreds of eighteenth- and nineteenth-century sightings of the descendants of Olaus's Sea Serpent (B d) failed to produce definitive identification of what was seen. Explanations have included suggestions as varied as basking sharks, oarfish, ribbonfish, eel, squid, a string of porpoises, seaweed, and logs. A saurian fossil sometimes compared to both the Great Sea Serpent and the Loch Ness Monster is the plesiosaur.

CARIBDIS / NORWEGIAN MAELSTRÖM

Olaus's *Carta Marina* Caribdis (B f) is perhaps the earliest map depiction of the Moskestraumen off the northwest coast of Norway. One of the world's strongest tidal currents, the eddy is dangerous to small craft but is not as deadly as Olaus or fantastic fiction presents it. The cycle of high and low tides twice daily creates conflicting currents that result in eddies.

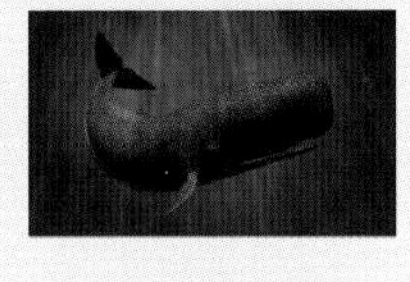

ANOTHER PRISTER / SPERM WHALE?

The sperm whale industry thrived during the eighteenth century and first half of the nineteenth, declined, and flourished again in the twentieth. Hunting of the species decimated the populations in different periods. The generally accepted view is that, after reduction of numbers in the twentieth century, the population of the now-protected sperm whale is slowly increasing, perhaps to about a million whales.

A SEA CREATURE / WALRUS OR SEAL?

Olaus's mermen derived from Pliny's Mediterranean Nereids and Tritons. Christopher Columbus's famous "mermaids" in the Caribbean are universally regarded as manatees, sirenian relatives of dugongs. Olaus's Greenland Sea Creature with the icy beard could correspond to a walrus with its bristly muzzle. The more mermaid-looking Sea Creature near the *Carta Marina*'s Island Whale (A l) might relate to the harbor seal.

ROSMARUS / WALRUS

The Atlantic Walrus's tusks, up to 3 feet in length, are even more pronounced than those of the *Carta Marina*'s monster with the "dreadful teeth" (B b). Up to 12 feet long and weighing as much as 1,640 pounds, the *Odobenus rosmarus rosmarus* is just as massive as Olaus's beast. Hunted to near extinction for its ivory, hide, and blubber, the walrus is now an endangered species.

KRAKEN / GIANT SQUID

The giant squid *(Architeuthis)* is the largest invertebrate on the earth. With eight arms and two long feeding tentacles extending from its head, it can grow to at least 60 feet in length and weigh more than a ton. Its eyes, up to 15 inches in diameter, are the largest of any animal. Japanese scientists were the first to photograph a living giant squid, in September 2004.

TIME LINE

Lore and Legacy of the Carta Marina

384–322 BC

Aristotle Greek philosopher, naturalist. *Historia Animalium* the West's first systematic natural history from observation.

AD 23–79

Pliny the Elder Roman compiler of authorities in the monumental *Natural History.* Olaus Magnus's principal classical source.

2nd–4th C.

Physiologus Christianized folklore of animals, basis of the bestiaries.

ca. 339–397

St. Ambrose Church Father. God's kingdom of animals in the *Hexameron.*

ca. 560–636

St. Isidore of Seville Church Father. Classical animal lore of the encyclopedic *Etymologies* expands the *Physiologus* into the medieval bestiaries.

11th–13th C.

Bestiaries Allegorized Christian "natural history" of both real and fabulous beasts. Evolved from the *Physiologus.*

ca. 1200–1280

Albertus Magnus German theologian, philosopher. *De Animalibus* is the most important zoological work between the natural histories of Pliny the Elder and Conrad Gesner.

1488–1544

Johannes Magnus Swedish ecclesiastic, historian. The last active Archbishop of Sweden. *History of all Kings of Goths and Swedes* (1554) published posthumously by his brother, Olaus.

1488–1552

Sebastian Münster German cartographer, cosmographer. *Monstra Marina & Terrestria,* derived from figures on Olaus's *Carta Marina,* in multiple editions of *Cosmographia* (1544).

1490–1557

Olaus Magnus Swedish ecclesiastic, cartographer, historian. Nominal Archbishop of Uppsala, Sweden, after his brother's death.
1539 *Carta Marina* "A Marine map and Description of the Northern Lands and of their Marvels, most carefully drawn up at Venice in the year 1539." Original copies are discovered in 1886 and 1962.
1555 *Historia de Gentibus Septentrionalibus* ("History of the Northern Peoples") is published in Rome. The authoritative work on Scandinavia for centuries after.

1491

Hortus Sanitatis An herbal and bestiary, with woodcuts. Dutch translation *The Palace of Animals* (1520).

1507–1566

Guillaume Rondelet French professor of medicine. A founder of modern marine biology. *Libri de Piscibus Marinis* (1554–1555).

1510–1590

Ambroise Paré French surgeon, regarded as the "father of surgery." Variations of *Carta Marina* sea monsters from Conrad Gesner in *Of Monsters and Prodigies.*

1512–1594

Gerard Mercator Flemish cartographer, creator of the Mercator Projection. Reproduced the *Carta Marina* sea cow and whale on the 1541 terrestrial globe.

1514–1587

Adriaen Coenen Dutch amateur naturalist. Figures and text derived from Olaus Magnus's sea monsters in his *Whale Book,* a 1585 manuscript not published until 2003.

1516–1565

Conrad Gesner Swiss naturalist, "the Father of Modern Zoology." *Historiae Animalium* (1551–1558) woodcuts from works of Pierre Belon, Guillaume Rondelet, and Olaus Magnus, reproduced in subsequent natural histories.

1517–1564

Pierre Belon French naturalist. A founder of modern marine biology. *Histoire naturelle des estranges poissons marins* (1551) and *De Aquatilibus Libri duo* (1553).

1527–1598

Abraham Ortelius Flemish cartographer. *Theatrum Orbis Terrarum* (1570), "the first world atlas." Sea monsters of *Islandia,* a map of Iceland, heavily influenced by Olaus's *Carta Marina.*

Resources

A substantial body of print and electronic material related to Olaus Magnus's *Carta Marina* and *Historia de Gentibus Septentrionalibus* has been produced since the early twentieth century. Such sources include the works described below:

One of the two surviving copies of Olaus Magnus's original 1539 *Carta Marina* is on display in the University of Uppsala Library, Sweden. The map, its panels, and its corresponding key are reproduced on the library's Web site. The University of Minnesota's James Ford Bell Library site presents a colored version of the *Carta Marina*, with pages of background information. The site and the supplemental article of the library curator Carol Urness, "Olaus Magnus: His Map and His Book," are indebted to the first and standard English study of Olaus's Scandinavian map: Edward Lynam's 1949 *The* Carta Marina *of Olaus Magnus*. Olaus Magnus scholar John Granlund reviews Lynam's book in his "The *Carta Marina* of Olaus Magnus." Herman Richter's 1967 *Olaus Magnus Carta marina 1539* is a rare source for the Italian and German versions of Olaus's *Carta Marina* key. Margareta Lindgren's lecture, "Olaus Magnus, Carta Marina of 1539, and Historia, 1555" is reproduced on the *Exploring Carta Marina* forum site.

The entire *Historia de Gentibus Septentrionalibus* is accessible in electronic format at Projekt Runeberg. The first English translation (abridged) of Olaus's work, *A Compendious History of the Goths, Swedes, and Vandals and Other Northern Nations* (1658), is on microfilm as well as in university libraries' special collections. The first complete English translation of Olaus's encyclopedic book is the definitive three-volume *A Description of the Northern Peoples 1555*, published by the Hakluyt Society (1996–98). The only other modern translation of the entire work is the Swedish version, published between 1909 and 1915. Vicki Ellen Szabo devotes a substantial section of her *Monstrous Fishes and the Mead-Dark Sea* to Olaus's treatment of Scandinavian marine life.

Works cited, consulted, or recommended for further reading

Among the distinguished modern works featuring sea monsters are those by the following authors. The classical *ketos*: John Boardman, Adrienne Mayor, John K. Papadopoulos and Deborah Ruscillo. Medieval sea beasts: Vicki Ellen Szabo; Renaissance marine life: Michon Scott. Nineteenth-century reassessment: John Ashton, Henry Lee, and A.C. Oudemans. And twentieth-century retrospectives: Cornelia Catlin Coulter, Richard Ellis, Bernard Heuvelmans, and Willy Ley.

The Aberdeen Bestiary. Translation and transcription by Colin McLaren. Aberdeen University Library. www.abdn.ac.uk/bestiary/bestiary.hti.

Aelian. *On Animals*. Vol. 2. Edited by E. H. Wormington. Loeb Classical Library, 1958. Reprint, Cambridge, Mass.: Harvard University Press, 1971.

Albertus Magnus. *Albert the Great: Man and the Beasts; de animalibus (Books 22–26)*. Translated by James J. Scanlan. Binghamton, N.Y.: Medieval & Renaissance Texts and Studies, 1987. See Book 24, "The Individual Aquatic Animals." James J. Scanlan's copious notes on medieval marine animals are invaluable.

Animal. Editors in Chief David Burnie and Don E. Wilson. Smithsonian Institution. New York: Dorling Kindersley, 2001.

"Arrow-Odd." In *Seven Viking Romances*. Translated by Hermann Palsson and Paul Edwards. New York: Viking Penguin, 1985.

Aristotle. *The History of Animals*. In *The Complete Works of Aristotle*. Vol. 1. Edited by Jonathan Barnes. Princeton, N.J.: Princeton University Press, 1984.

Arrian. *History of Alexander and Indica*. Vol. 2. Translated by P. A. Brunt. Loeb Classical Library. Cambridge, Mass.: Harvard University Press, 1983.

Ashton, John. *Curious Creatures in Zoology*. London, 1890. Facsimile ed. Google books, 2012.

Babylonian Talmud. Translated by Michael L. Rodkinson, 1918. Internet Sacred Text Archive. www.sacred-texts.com.

Barber, Richard. *Bestiary*. Woodbridge, Suffolk: Boydell Press, 1999. Translation of MS Bodley 764.

Bartholomew Anglicus. In Robert Steele. *Mediaeval Lore from Bartholomew Anglicus*. New York: Cooper Square, 1966.

Black, Jeremy, and Anthony Green. *Gods, Demons and Symbols of Ancient Mesopotamia*. 1992. Reprint, Austin: University of Texas Press, 2000.

Blefken, Dithmar. *Islandia* (1607). In *Purchas, Hakluytus posthumus: or Purchas his Pilgrimes*. Vol. 13. Glasgow: James MacLehose, 1905–1907.

Boardman, John. "'Very Like a Whale'—Classical Sea Monsters." In *Monsters and Demons in the Ancient and Medieval Worlds*. Edited by Ann E. Farkas, Prudence O. Harper, and Evelyn B. Harrison. Mainz on Rhine.

Verlag, 1987. The standard study of sea beasts in Greco-Roman art.

The Book of the Settlement of Iceland (Landnámabók). New Northvegr Center. www.northvegr.org.

The Book of the Thousand and One Nights. Translated and edited by Richard F. Burton. 1934. Vol. 3. Reprint, New York: Heritage Press, 1962.

Browne, Sir Thomas. *Pseudodoxia Epidemica* ("Vulgar Errors," 1672). penelope.uchicago.edu/pseudodoxia/pseudodoxia.shtml.

Carwardine, Mark. *Whales, Dolphins and Porpoises.* New York: Dorling Kindersley, 2002.

Chase, Owen. *Narrative of the Most Extraordinary and Distressing Shipwreck of the Whale-ship Essex.* New York: W. B. Gilley, 1821. MobileRead. www.mobileread.com.

Coenen, Adriaen. *The Whale Book: Whales and other marine animals as described by Adriaen Coenen in 1585.* Edited by Florike Egmond and Peter Mason with commentaries by Kees Lankester. London: Reaktion, 2003. An important amateur contribution to Renaissance natural history.

Coulter, Cornelia Catlin. "The 'Great Fish' in Ancient and Medieval Story," *Transactions and Proceedings of the American Philological Association*, 57 (1926): 32–50. A classic transmission study.

Ctesias. *Indica.* In *Ancient India as Described by Ktesias the Knidian*, edited by J. W. McCrindle. 1882. Reprint, Delhi, India: Manohar Reprints, 1973.

De Montfort, Pierre Denys. *Histoire naturelle générale et particulière des mollusques.* Paris, 1802. Google Books. books.google.com.

Der dieren palleys ("The Palace of Animals"). www.let.ru.nl/dierenpalleys/index.php?title (last accessed June 8, 2011; no longer available).

Druce, George C. "On the Legend of the Serra or Saw-Fish." *Proceedings of the Society of Antiquaries*, Second Series, Vol. 31. London: 1919.

Dunn, Joseph. "The Brendan Problem," *The Catholic Historical Review* 6:4 (Jan., 1921): 395–477.

Egede, Hans. *A Description of Greenland.* London: 1818. Google Books. books.google.com.

Ehrensvärd, Ulla. *The History of the Nordic Map: From Myths to Reality.* Translated by Roy Hodson. Helsinki: John Nurminen Foundation, 2006. The definitive work on the subject.

Ellis, Richard. *The Book of Whales.* New York: Knopf, 1980.
———. *Encyclopedia of the Sea.* New York: Knopf, 2006.
———. *Monsters of the Sea.* 1995. Reprint, New York: Lyons Press, 2001.
———. *The Search for the Giant Squid.* 1998. Reprint, New York: Penguin, 1999.

"The Epic of Creation." In *Myths from Mesopotamia.* Translated by Stephanie Dalley. New York: Oxford University Press, 1991.

Fournival, Richard. *Master Richard's Bestiary of Love and Response.* Translated by Jeanette Beer. Berkeley: University of California Press, 1986.

George, Wilma. *Animals and Maps.* Berkeley: University of California Press, 1969.

Gerard, John. *Gerard's Herball.* 1636. Edited by Marcus Woodward. London: Spring Books, 1964.

Gesner, Konrad. *Curious Woodcuts of Fanciful and Real Beasts.* New York: Dover, 1971.
———. *Nomenclator aquatilium animantium. Icones animalium aquatilium.* Harvard University Library Page Delivery Service. pds.lib.harvard.edu/pds/.

Godwin, Joscelyn. *Athanasius Kircher's Theatre of the World.* London: Thames & Hudson, 2009.

Granlund, John. "The *Carta Marina* of Olaus Magnus." *Imago Mundi* 8 (1951): 35–43.

Guthrie, William. *Geographical, Historical, and Commercial Grammar.* London, 1801. Google Books. books.google.com.

Heron-Allen, Edward. *Barnacles in Nature and in Myth.* London, 1878. Facsimile ed. Kessinger Publishing, n.d. The standard study of the subject.

Heuvelmans, Bernard. *In the Wake of the Sea-Serpents.* New York: Hill & Wang, 1968.

Hoare, Philip. *The Whale: In Search of the Giants of the Sea.* New York: Ecco, 2010.

Holman, Louis A. *Old Maps and Their Makers.* Boston: Charles E. Goodspeed, 1926.

Homer. *The Odyssey.* Translated by Samuel Butler. 1900. The Internet Classics Archive. classics.mit.edu/Homer/odyssey.12.xii.html.

Hortus Sanitatis ("The Garden of Health"). 1491. Digitale Bibliothek—Münchener Digitalisierungszentrum. dfg-viewer.de.

Isidore of Seville. *The* Etymologies *of Isidore of Seville.* Translated by Stephen A. Barney, W. J. Lewis, J. A. Beach, and Oliver Berghof. New York: Cambridge University Press, 2006. For fish, see Book 12, Chapter 6.

Johannesson, Kurt. *The Renaissance of the Goths in Sixteenth-Century Sweden: Johannes and Olaus Magnus as Politicians and Historians.* Berkeley: University of California Press, 1991.

The King's Mirror. Translated by Laurence Marcellus. New York: Twayne, 1917.

Lee, Henry. *Sea Monsters Unmasked.* London: 1883. Facsimile ed. Forgotten Books, n.d.

"Letter to Aristotle." In Pseudo-Callisthenes. *The Life of Alexander of Macedon.* Edited and translated by Elizabeth Hazelton Haight. New York: Longmans, Green, 1955.

Ley, Willy. *Dawn of Zoology.* Englewood Cliffs, N.J.: Prentice-Hall, 1968.
———. *Willy Ley's Exotic Zoology.* New York: Viking Press, 1959.

Lindgren, Margareta. "Olaus Magnus, Carta Marina of 1539, and Historia, 1555." *Exploring Carta Marina.* cipher.uiah.fi/forum/events/one_event/speech?lang=en&ex=ml.

Lucian. "A True Story." In *Lucian.* Vol. 1. Translated by A. M. Harmon. 1913. Reprint, Loeb Classical Library. Cambridge, Mass.: Harvard University Press, 1991.

Lynam, Edward. *The* Carta Marina *of Olaus Magnus, Venice 1539 & Rome 1572.* Jenkintown, Pa.: Tall Tree Library, 1949.

Magnus, Olaus. "Carta Marina." Uppsala University Library. www.ub.uu.se/en/Collections/Map-collections/Section-for-Maps-and-Pictures-map-collection/Carta-Marina/. The digitized *Carta Marina,* with panels and keys.
———. *A Compendious History of the Goths, Swedes & Vandals and Other Northern Nations.* London, 1658. Ann Arbor, Mich.: University Microfilms, Early English Books.
———. *A Description of the Northern Peoples 1555.* 3 vols. Edited by Peter Foote. Translated by Peter Fisher and Humphrey Higgens. Annotation derived from the Commentary of John Granlund. London: The Hakluyt Society, 1996–98. John Granlund's extensive annotation of Olaus's vignettes, sources, and text are indispensable.
———. *Historia de Gentibus Septentrionalibus.* Projekt Runeberg. runeberg.org/olmagnus/.

Mandeville, John. *The Travels of Sir John Mandeville.* Reprint, New York: Dover, n.d.

"Marvels in Marine Natural History." In *Colburn's United Service Magazine.* Part 1. London, 1846. Sea monster lore, pages 161–172 and 331–342. Facsimile ed. Google books, 2012.

Mayor, Adrienne. "Sea Monsters and Other Ancient Beasts: A Tale from a Grecian Urn." *Archaeology Odyssey* (March/April 2002): 44–52. The *ketos* in classical Greek art.

Mead, William R. "Scandinavian Renaissance Cartography." In *The History of Cartography.* Vol. 3,

Part 2. Edited by David Woodward. Chicago, Ill.: University of Chicago Press, 2007.

Melville, Herman. *Moby-Dick or, The Whale*. 1851. Reprint, New York: Modern Library, 2000.

Münster, Sebastian. *A Briefe Collection and compendious extract of strange and memorable thinges, gathered oute of the Cosmographye of Sebastian Munster*. London, 1572. The Library of Congress.

———. *Cosmographia*. Antwerp, 1584. The Library of Congress. The work's Latin key of the *Monstra Marina & Terrestria* plate was translated for this book by Mary Margolies DeForest, Denver, Colo., 2011.

Nigg, Joseph. *The Book of Fabulous Beasts: A Treasury of Writings from Ancient Times to the Present*. New York: Oxford University Press, 1999.

Ogilvie, Brian W. *The Science of Describing: Natural History in Renaissance Europe*. 2006. Reprint, Chicago, Ill.: University of Chicago Press, 2008.

Ortelius, Abraham. "Cartographica Neerlandica Map Text for Ortelius Map No. 161." *Cartographica Neerlandica*. www.orteliusmaps.com/book/ort_text161.html.

Oudemans, A. C. *The Great Sea Serpent*. 1892. Facsimile ed. New York: Cosimo Classics, 2009.

Papadopoulos, John K., and Deborah Ruscillo. "A Ketos in Early Athens: An Archaeology of Whales and Sea Monsters in the Greek World." *American Journal of Archaeology* 106.2 (April 2002): 187–227. An essential study of classical sea beasts.

Paré, Ambroise. *Of Monsters and Prodigies*. In *The Collected Works of Ambroise Paré*. Translated by Thomas Johnson. 1634. Facsimile ed. Pound Ridge, N.Y.: Milford House, 1968.

Payne, Anne. *Medieval Beasts*. London: The British Library, 1989.

Philippe de Taun. *The Bestiary*. In *Popular Treatises on Science*. Edited by Thomas Wright. London, 1841.

Physiologus. Translated by Michael J. Curley. Austin: University of Texas Press, 1979.

Pliny the Elder. *Natural History*. Vol. 3, others. Translated by H. Rackham. Loeb Classical Library, 1940. Reprint, Cambridge, Mass.: Harvard University Press, 1983. See Book 9, on fishes and sea monsters.

Poe, Edgar Allan. "A Descent into the Maelström." In *The Works of Edgar Allan Poe*. Vol. 3. New York: Funk & Wagnalls, 1904.

Polo, Marco. *The Travels of Marco Polo, The Complete Yule-Cordier Edition*. Vol. 2. Edited by Sir Henry Yule and Henri Cordier. 1903. Reprint, New York: Dover, 1993.

Pontoppidan, Erich. *Natural History of Norway*. Part 2. London, 1755. Facsimile edition, Ecco, n.d.

Rabelais, François. *Gargantua and His Son Pantagruel*. Translated by Sir Thomas Urquhart and Peter Antony Motteux. Derby: Moray Press, 1894. Project Gutenberg. www.gutenberg.org/files/8169/8169-h/8169-h.htm.

Richter, Herman. *Olaus Magnus Carta marina 1539*. Lund, Sweden: Lychnos-Bibliotek, 1967.

Røde, Grø, and Otta Schiøotz. *Guide to the History of Lofoten*. Lofoten Public Museum/Lofoten Regional Council, 1996. *Lofoten Islands*. www.lofoten-info.no/history.htm.

Rossby, H. Thomas, and Peter Miller. "Ocean Eddies in the 1539 Carta Marina by Olaus Magnus." *Oceanography* 16.4, 2003: 77–88.

Scott, Michon. "Sea Monsters." *Strange Science*. www.strangescience.net/stsea2.htm. The most comprehensive online survey of Renaissance sea monsters.

Scott, Sir Walter. *The Pirate*. 1822. In *The Waverley Novels*. Vol. 13. Reprint, New York: P. F. Collier & Son, 1902.

Shepard, Odell. *The Lore of the Unicorn*. 1930. Reprint, New York: Avenel Books, 1982.

St. Ambrose. *Saint Ambrose: Hexameron, Paradise, and Cain and Abel*. Translated by John J. Savage. New York: Fathers of the Church, 1961. For fishes, see Book 5, "The Fifth Day."

Strabo. *Geography*. Vols. 1, 3, and 7. Edited by Horace Leonard Jones. Loeb Classical Library, 1930. Reprint, Cambridge, Mass.: Harvard University Press, 1995.

Sturluson, Snorri. *The Prose Edda*. Translated by Arthur Gilchrist Brodeur. New York: American-Scandinavian Foundation, 1960.

Swan, John. *Speculum Mundi: or, A Glasse Representing the Face of the World*. London, 1665. Ann Arbor, Mich.: University Microfilms, Early English Books.

Sweeney, James B. *A Pictorial History of Sea Monsters and Other Dangerous Marine Life*. New York: Crown, 1972.

Szabo, Vicki Ellen. *Monstrous Fishes and the Mead-Dark Sea*. Leiden: Brill, 2008. The comprehensive standard study of medieval marine life in the northern seas.

Topsell, Edward. *The History of Four-Footed Beasts and Serpents and Insects*. Vol. 2. London, 1658. Introduction by Willy Ley. Facsimile ed. New York: Da Capo Press, 1967.

Tooley, R. V. *Maps and Map-Makers*. 1949. Reprint, New York: Crown, 1982.

Urness, Carol. "Olaus Magnus: His Map and His Book." *Mercator's World* 6 no. 1 (2001): 26–33.

Verne, Jules. *Twenty Thousand Leagues Under the Seas*. Boston: G. M. Smith, 1874. MobileRead. www.mobileread.com.

Westenholz, Joan Goodnick. *Dragons, Monsters and Fabulous Beasts*. Jerusalem: Bible Lands Museum, 2004.

"Whale." In *Early English Christian Poetry*. Translated and edited by Charles W. Kennedy. 1952. Reprint, New York: Oxford University Press, 1968.

Whitbourne, Richard. "Captaine Richard Whitbournes voyages to New-found-land." In Samuel Purchas, *Hakluytus Posthumus, or Purchas His Pilgrimes*. Vol. 7. Glasgow: James MacLehose, 1905–1907.

White, T. H. *The Book of Beasts*. 1954. Reprint, New York: Dover, 1984.

Whitfield, Peter. *The Charting of the Oceans: Ten Centuries of Maritime Maps*. London: The British Library, 1996.

Wittkower, Rudolf. *Allegory and the Migration of Symbols*. London: Thames and Hudson, 1977.

Wonders of the Deep: The Astonishing Splendor of the Seven Seas. New York: Life Books, June 2012.

Web sites

Bibliodyssey
bibliodyssey.blogspot.com/
Images from rare books and prints, including Renaissance natural histories.

Exploring Carta Marina
cipher.uiah.fi/forum/events/one_event/speech?lang=en&ex=ml
A Nordic heritage forum based on Olaus Magnus's 1539 map. Sponsored by CIPHER.

"Marine Biology Before Linnaeus." Coasts and Reefs.
www.coastsandreefs.net/books/timeline/index.html
A survey of major naturalists from Aristotle to Carolus Linnaeus.

The Medieval Bestiary
bestiary.ca/
Background, links to online bestiaries.

"Olaus Magnus' Scandinavia." James Ford Bell Library, University of Minnesota.
www.lib.umn.edu/apps/bell/map/OLAUS/indexo.html
Background of Olaus Magnus and a color reproduction of the *Carta Marina* and its individual panels.

Service Commun de la Documentation de l'Université de Strasbourg
docnum.u-strasbg.fr/cdm4/browse.php?CISOROOT=%2Fcoll13
Contains digitized Renaissance natural histories.

Index

~ N ~

~ O ~

~ P ~

~ R ~

~ S ~

~ T ~

~ U ~

~ V ~

~ W ~

~ Z ~

ACKNOWLEDGMENTS

Author Acknowledgments

I express my deep appreciation to the many individuals who contributed to the making of this book.

Sea Monsters owes its form to the conception of Jason Hook, the publisher of Ivy Press. He envisioned my voyage idea in terms of iconic *Carta Marina* images, supplemented with background text. Ivy editorial director Caroline Earle expertly guided the complex production process, transforming art and text through the combined talents of art director Michael Whitehead, designer Andrew Milne, and project editor Jamie Pumfrey. The result gives Olaus's centuries-old sea monsters the artful attention they deserve.

I extend special thanks to the following: scholarly readers Adrienne Mayor and Victoria Morse for their insightful suggestions and bibliographical sources; Mary Margolies DeForest for her translation of Sebastian Münster's chart key; Marcel van den Broecke (Cartographical Neerlandica) for the *Islandia* key and information; Christopher W. Lane (The Philadelphia Print Shop West), Curtis Bird (The Old Map Gallery), and map collector Jeff Miller for their cartographical knowledge; Michon Scott for Renaissance natural history sources; Joseph Hutchison for technical assistance; naturalist Jim Nelson for proofreading; and Lawrence Dunning, Dr. Richard Hagman, Marjorie Muzzillo, Cindy Wold, Cassidy Nigg, and Oliver Monk for their interest in the book.

The spirit of my elder son, Joey (Joseph Conrad Nigg), and my father (Joseph John Nigg, May 13, 1907–April 30, 1970) were inspiration throughout the Voyage. And, as always, I specially thank my wife, Esther. She not only shared with me her books, classical and Renaissance knowledge, and linguistic skills, but also provided unlimited emotional support from the first day of this project to the last.

Joseph Nigg

The Ivy Press would like to thank the following for permission to reproduce or adapt copyright material:

Marcel and Deborah van den Broecke (Cartographica Neerlandica, www.orteliusmaps.com) for permission to adapt their translation of the *Islandia* map key: 17.

Mary Margolies DeForest for permission to adapt her translation of *Monstra Marina & Terrestria* map key: 14.

Uppsala University Library for permission to adapt their translation of the full *Carta Marina* map key: 150–151.

Image Credits

AKG Images: 54B, 76B, 97T; British Library: 20, 108, 132T; British Museum: 139; De Agostini Picture Library: 11; Florilegius: 81; historic-maps: 102B. **Alamy/The Art Archive:** 32; North Wind Picture Archives: 147T; Photos12: 122B; WildlifeGmbH: 90T. **Bayerische Staatsbibliothek München** [Res/2 M.med. 45d.]: 76C, 128T. **Bridgeman Art Library/ Arni Magnusson Institute, Reykjavik, Iceland:** 116. **Corbis:** 64C; Blue Lantern Studio: 123B; Bettmann: 15, 48B, 54T, 61T, 70T, 76T, 86T, 102T, 109, 117T, 129B, 145T, 147B; Stefano Bianchetti: 146T; Fine Art Photographic Library: 16, 64B, 111T, 138B; The Gallery Collection: 19; Time Life Pictures/ Mansell: 91T; Underwood & Underwood: 122T. **Fotolia**: 153; Bernd Ege: 133B. **Getty Images/The Bridgeman Art Library:** 19; Hulton Archive: 118; Universal Images Group: 110. **James Ford Bell Library, University of Minnesota:** 5, 13, 22, 25, 27TR, 28BR, 29, 34BR, 35, 40BR, 41, 44BR, 45, 50BR, 51, 56BR, 57, 62BR, 63, 66BR, 67, 72BR, 73, 78, 79BR, 80T, 82BR, 83, 88BR, 89, 90B, 92BR, 93, 98BR, 99, 104, 105, 112BR, 113, 120BR, 121, 124BR, 125, 130, 131, 134BR, 135, 140BR, 141, 148TL **Librairie Droz:** 33TR, 60T. **Library of Congress:** 119B, 119T, 154. **Library of the University of Seville:** 70B, 103B. **Mary Evans Picture Library:** 86B; Interfoto/Sammlung Rauch: 80. **Museum of Antiquities, University of Saskatchewan/Kathryn Brooks:** 144B. **Royal Zoological Society of Antwerp:** 39, 49T, 55, 61B, 87, 103T, 111B, 128B, 144T. **Scala Archives/The Pierpont Morgan Library/Art Resource:** 77. **Science Photo Library/Paul D Stewart:** 38. **Shutterstock:** 152, 153; Brendan Howard: 71T, Gail Johnson: 152CR; Hugh Lansdown: 152BR; Morphart Creation: 49B; Christian Musat: 152L; Nicku: 18; Picturepartners: 152CL; Edward Westmacott: 152BL. **Thinkstockphoto:** 153. **Topfoto/British Library Board/Robana:** 42T; The Granger Collection: 42, 43. **Uppsala University Library:** 8C.

Every effort has been made to trace copyright holders and to obtain their permission for the use of copyright material. The publisher apologizes for any errors or omissions in the lists above and will gratefully incorporate any corrections in future reprints if notified.